BMW
Automotive Repair Manual

by John H Haynes
Member of the Guild of Motoring Writers

and Peter Strasman MISTC

Models covered:

All versions (two & three door) of the following:
BMW 1500. 1499cc (91.4 cu in)
BMW 1502. 1573cc (95.9 cu in)
BMW 1600. 1573cc (95.9 cu in)
BMW 1602. 1573cc (95.9 cu in)
BMW 2000 Touring. 1990cc (121.4 cu in)
BMW 2002 (except 'Turbo'). 1990cc (121.4 cu in)

ISBN 0 85696 240 6

Haynes UK (18050-6W5)
Sparkford Nr Yeovil
Somerset BA22 7JJ England

Haynes North America, Inc
859 Lawrence Drive
Newbury Park, CA 91320 USA

www.haynes.com

Acknowledgements

We are indebted to Bayerische Motorenwerke A.G. (BMW) for their assistance in the provision of technical information and for permission to reproduce certain of their illustrations. Castrol Limited supplied lubrication data.

Special thanks are due to Wyke Regis Garage, Weymouth, Dorset and their associate company Camel Cross Motors, West Camel, Yeovil, Somerset, for loaning us the BMW 2002 which was used as our project car. Mark Fowler and Dave Napper were particularly helpful.

Many of the photographs used in the bodywork repair section of Chapter 12, were supplied by *Car Mechanics* magazine.

Lastly we are grateful to all of those people at Sparkford who helped in the production of this manual. Particularly Brian Horsfall and Les Brazier, who carried out the mechanical work and took the photographs respectively; Les Burch who planned the layout of each page, and Rod Grainger the editor.

Introduction to the BMW

The BMW models covered by this manual are soundly constructed and mechanical components are engineered to fine limits. A buyer contemplating the purchase of one of these cars will be reassured by the knowledge that they are absolutely conventional in design and should cause no problems in overhaul or repair.

The fact that the series has enjoyed such long production runs over many years provides a hedge against depreciation and ensures the availability of both new and second-hand spare parts.

All models described in this manual are instantly identified by having only two door bodywork with the exception of the 2000 Touring which has a tailgate (hatchback) to give it a semi-estate appearance.

About this manual

Its aims

The aim of this book is to help you get the best value from your car. It can do so in two ways. First it can help you decide what work must be done, even should you choose to get it done by a garage, the routine maintenance and the diagnosis and course of action when random faults occur. But it is hoped that you will also use the second and fuller purpose by tackling the work yourself. This can give you the satisfaction of doing the job yourself. On the simpler jobs it may even be quicker than booking the car into a garage and going there twice, to leave and collect it. Perhaps most important, much money can be saved by avoiding the costs a garage must charge to cover their labour and overheads.

The book has drawings and descriptions to show the function of the various components so that their layout can be understood. Then the tasks are described and photographed in a step-by-step sequence so that even a novice can cope with complicated work.

The jobs are described assuming only normal tools are available, and not special tools. But a reasonable outfit of tools will be a worthwhile investment. Many special workshop tools produced by the makers merely speed the work, and in these cases guidance is given as to how to do the job without them. On a very few occasions the special tool is essential to prevent damage to components, then their use is described. Though it might be possible to borrow the tool, such work may have to be entrusted to the official agent.

Using the manual

The book is divided into twelve Chapters. Each Chapter is divided into numbered Sections which are headed in **bold type** between horizontal lines. Each Section consists of serially numbered paragraphs.

There are two types of illustration: (1) Figures which are numbered according to Chapter and sequence of occurrence in that Chapter. (2) Photographs which have a reference number on their caption. All photographs apply to the Chapter in which they occur so that the reference figure pinpoints the pertinent Section and paragraph number.

Procedures, once described in the text, are not normally repeated. If it is necessary to refer to another Chapter the reference will be given in Chapter number and Section number thus: Chapter 1/16.

If it is considered necessary to refer to a particular paragraph in another Chapter the reference is given in this form: 1/5:5. Cross-references given without use of the word 'Chapter' apply to Sections and/or paragraphs in the same Chapter (eg; 'see Section 8') means also 'in this Chapter'.

When the left or right side of the car is mentioned it is as if looking forward from the rear of the car.

Great effort has been made to ensure that this book is complete and up-to-date. However, it should be realised that manufacturers continually modify their cars, even in retrospect.

Whilst every care is taken to ensure that the information in this manual is correct no liability can be accepted by the authors or publishers for loss, damage or injury caused by any errors in, or omissions from, the information given.

Contents

BMW 1602 - 1974 model (Inset: BMW 1502 - 1975 model)

BMW 2002 - 1974 model (Inset: BMW 2002 Touring - 1974 model)

Use of English

As this book has been written in England, it uses the appropriate English component names, phrases, and spelling. Some of these differ from those used in America. Normally, these cause no difficulty, but to make sure, a glossary is printed below. In ordering spare parts remember the parts list may use some of these words:

English	American	English	American
Accelerator	Gas pedal	Locks	Latches
Aerial	Antenna	Methylated spirit	Denatured alcohol
Anti-roll bar	Stabiliser or sway bar	Motorway	Freeway, turnpike etc
Big-end bearing	Rod bearing	Number plate	License plate
Bonnet (engine cover)	Hood	Paraffin	Kerosene
Boot (luggage compartment)	Trunk	Petrol	Gasoline (gas)
Bulkhead	Firewall	Petrol tank	Gas tank
Bush	Bushing	'Pinking'	'Pinging'
Cam follower or tappet	Valve lifter or tappet	Prise (force apart)	Pry
Carburettor	Carburetor	Propeller shaft	Driveshaft
Catch	Latch	Quarterlight	Quarter window
Choke/venturi	Barrel	Retread	Recap
Circlip	Snap-ring	Reverse	Back-up
Clearance	Lash	Rocker cover	Valve cover
Crownwheel	Ring gear (of differential)	Saloon	Sedan
Damper	Shock absorber, shock	Seized	Frozen
Disc (brake)	Rotor/disk	Sidelight	Parking light
Distance piece	Spacer	Silencer	Muffler
Drop arm	Pitman arm	Sill panel (beneath doors)	Rocker panel
Drop head coupe	Convertible	Small end, little end	Piston pin or wrist pin
Dynamo	Generator (DC)	Spanner	Wrench
Earth (electrical)	Ground	Split cotter (for valve spring cap)	Lock (for valve spring retainer)
Engineer's blue	Prussian blue	Split pin	Cotter pin
Estate car	Station wagon	Steering arm	Spindle arm
Exhaust manifold	Header	Sump	Oil pan
Fault finding/diagnosis	Troubleshooting	Swarf	Metal chips or debris
Float chamber	Float bowl	Tab washer	Tang or lock
Free-play	Lash	Tappet	Valve lifter
Freewheel	Coast	Thrust bearing	Throw-out bearing
Gearbox	Transmission	Top gear	High
Gearchange	Shift	Torch	Flashlight
Grub screw	Setscrew, Allen screw	Trackrod (of steering)	Tie-rod (or connecting rod)
Gudgeon pin	Piston pin or wrist pin	Trailing shoe (of brake)	Secondary shoe
Halfshaft	Axleshaft	Transmission	Whole drive line
Handbrake	Parking brake	Tyre	Tire
Hood	Soft top	Van	Panel wagon/van
Hot spot	Heat riser	Vice	Vise
Indicator	Turn signal	Wheel nut	Lug nut
Interior light	Dome lamp	Windscreen	Windshield
Layshaft (of gearbox)	Countershaft	Wing/mudguard	Fender
Leading shoe (of brake)	Primary shoe		

Buying spare parts and vehicle identification numbers

Buying spare parts

Replacement parts are available from many sources, which generally fall into one of two categories – authorized dealer parts departments and independent retail auto parts stores. Our advice concerning these parts is as follows:

Retail auto parts stores: Good auto parts stores will stock frequently needed components which wear out relatively fast, such as clutch components, exhaust systems, brake parts, tune-up parts, etc. These stores often supply new or reconditioned parts on an exchange basis, which can save a considerable amount of money. Discount auto parts stores are often very good places to buy materials and parts needed for general vehicle maintenance such as oil, grease, filters, spark plugs, belts, touch-up paint, bulbs, etc. They also usually sell tools and general accessories, have convenient hours, charge lower prices and can often be found not far from home.

Authorized dealer parts department: This is the best source for parts which are unique to the vehicle and not generally available elsewhere (such as major engine parts, transmission parts, trim pieces, etc.).

Warranty information: If the vehicle is still covered under warranty, be sure that any replacement parts purchased – regardless of the source – do not invalidate the warranty!

To be sure of obtaining the correct parts, have engine and chassis numbers available and, if possible, take the old parts along for positive identification.

Vehicle identification numbers

Modifications are a continuing and unpublicised process in vehicle manufacture quite apart from major model changes. Spare parts manuals and lists are compiled upon a numerical basis, the individual vehicle number being essential to correct identification of the component required.

The *engine number* is located on the crankcase just above the starter motor.

The *vehicle identification plate* is located within the engine compartment on the side wheel arch.

O*n North American* cars, the *vehicle identification number* is repeated on the top surface of the instrument panel just inside the windshield.

Engine number

Routine maintenance

Maintenance is essential for ensuring safety and desirable for the purpose of getting the best in terms of performance and economy from the car. Over the years the need for periodic lubrication - oiling, greasing and so on - has been drastically reduced if not totally eliminated. This has unfortunately tended to lead some owners to think that because no such action is required the items either no longer exist or will last for ever. This is a serious delusion. It follows therefore that the largest initial element of maintenance is visual examination. This may lead to repairs or renewals.

In the summary given here the essential for safety items are shown in **bold type**. These **must** be attended to at the regular frequencies shown in order to avoid the possibility of accidents and loss of life. Other neglect results in unreliability, increased running costs, more rapid wear and more rapid depreciation of the vehicle in general.

Every 250 miles (400 km) or weekly

Steering
Check the tyre pressures, including the spare
Examine tyres for wear or damage
Check steering for slackness

Brakes
Check reservoir fluid level
Check brake pedal travel
Check efficiency of handbrake

Electrical
Check all lights
Check washer and wiper efficiency and washer jet adjustment

Engine
Check oil level
Check coolant level
Check battery electrolyte level

At first 1000 miles (1600 km)

With new cars, it is recommended that the engine oil, gearbox oil and final drive oils are changed. Drain the oils when warm and also renew the engine oil filter.

Every 5000 miles (80 00 km)

Steering
Check linkage, balljoints and bushes for wear (lubricate balljoints on very early models only)
Check and adjust if necessary, front hub bearings
Check steering box oil level
Check security of steering box mounting bolts

Brakes
Check pad and lining wear
Adjust rear brakes
Check fluid lines for corrosion and leaks

Suspension
Check all mountings, pivots and bushes for wear
Check hydraulic shock absorbers for signs of leakage

Engine
Change engine oil and renew filter
Check distributor points gap and timing
Clean spark plugs and re-gap

Transmission and final drive
Check oil level and top-up, if necessary
Grease driveshaft inner universal joints (needle bearing type only)

Body
Lubricate all locks and hinges

Every 10000 miles (16000 km)

Steering
Check front wheel alignment
Re-balance roadwheels and tyres

Clutch
Adjust free-movement (where applicable)

Engine

Check carburettor adjustment
Check fan belt tension
Check cylinder head bolt torque wrench settings
Check and adjust valve clearances
Renew air cleaner element
Clean fuel filters and pump (carburettor models)
Renew spark plugs
Renew distributor points

Every 20000 miles (32000 km)

Engine

Clean EGR (emission control) system pipes
Check compression for evaluation of engine condition

Transmission

Check propeller shaft joints for wear
Change manual gearbox or automatic transmission oil
Renew oil in driveshaft outer joints (sliding type only)

Every 40000 miles (64000 km)

Engine

Clean fuel pump filter and renew main line filter (2002 TII only)

Brakes

**Drain hydraulic fluid and renew all seals in master and wheel cylinders.
Fill with fresh fluid and bleed**

Hubs

Dismantle, clean, repack with lubricant and adjust front and rear hub bearings

Every spring and autumn

Turn pre-heat valve lever to seasonal position

Every two years

Drain and flush the cooling system and refill with antifreeze mixture

Additional items shich should be attended to, when the time can be spared

Cleaning

Examination of components requires that they be cleaned. The same applies to the body of the car, inside and out, in order that deterioration due to rust or unknown damage may be detected. Certain parts of the body frame, if rusted badly, can result in the vehicle being declared unsafe and it will not pass a test for roadworthiness.

Exhaust system

An exhaust system must be leakproof, and the noise level below a certain minimum. Excessive leaks may cause carbon monoxide fumes to enter the passenger compartment. Excessive noise constitutes a public nuisance. Both these faults may cause the vehicle to be kept off the road. Repair or replace defective sections when symptoms are apparent.

Location of spare wheel and tools.

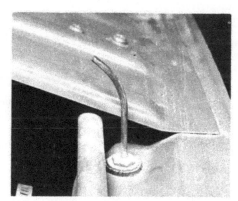
Windscreen washer jet.

Topping-up the engine oil.

Gearbox filler plug.

Topping-up the gearbox oil

Gearbox drain plug.

General dimensions and capacities

	1500/1600	1502/1602	2002	2000 Touring
Overall length	177.2 in (450.0 cm)	166.5 in (422.9 cm)	166.5 in (422.9 cm)	161.8 in (410.9 cm)
Overall width	67.3 in (171.0 cm)	62.6 in (159.0 cm)	62.6 in (159.0 cm)	62.6 in (159.0 cm)
Overall height	57.1 in (145.0 cm)	55.5 in (141.0 cm)	55.5 in (141.0 cm)	54.3 in (138.0 cm)
Ground clearance	5.9 in (150.0 mm)	6.25 in (159.4 mm)	6.25 in (159.4 mm)	6.5 in (165.1 mm)
Wheelbase	100.0 in (254.0 cm)	98.4 in (250.0 cm)	98.4 in (250.0 cm)	98.4 in (250.0 cm)
Unladen weight	2010 lb (912 kg)	2072 lb (940 kg)	2072 lb (940 kg)	2271 lb (1030 kg)
			2002 TII: 2183 lb (990 kg)	

Capacities

Engine oil (renewing filter)	7.5 Imp. pints; 4.25 litres; 9.03 US pints
Cooling system	12.3 Imp. pints; 7.0 litres; 7.4 US quarts
Manual gearbox:	
Long extension type	2.20 Imp. pints 1.25 litres 2.64 US pints
Universal 232 type and 5 speed type	2.52 Imp. pints; 1.4 litres; 2.94 US pints
Automatic transmission	8.2 Imp. pints; 4.65 litres; 4.9 US quarts
Final drive	1.6 Imp. pints; 0.9 litres; 0.95 US quarts
Fuel tank:	
1500/1600 models	12.1 Imp. gallons; 55.0 litres; 14.5 US gallons
Other models	10.1 Imp. gallons; 46.0 litres; 12.2 US gallons

Jacking and towing points

The jack supplied with the car tool kit should only be used for changing roadwheels. When carrying out repairs to the car, jack-up the front end under the crossmember or the rear under the differential/final drive unit. Always supplement these jacks with axle stands placed under the bodyframe members.

A towing eye is provided at the front and rear of the car for emergency use, but do not tow vehicles which are of considerably larger size and weight than your own. If your car is equipped with automatic transmission, it must only be towed if the speed selector lever is first placed in 'N' and the road speed must not exceed 30 mph (48 kph). The total distance covered must not exceed 30 miles (48 km) unless an additional 2 Imp. pints (1.1 litres) of fluid are added to the transmission unit or the propeller shaft disconnected, otherwise lack of lubrication may damage the internal components. Always reduce the fluid level again when the car is ready to operate normally. Tow starting or starting the car by running it down an incline is not possible with the type of automatic transmission fitted.

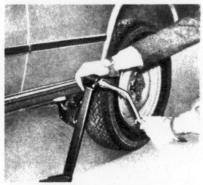

Using jack supplied in car tool kit

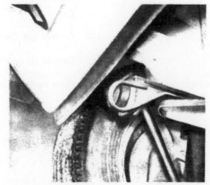

Front towing eye

Rear towing eye

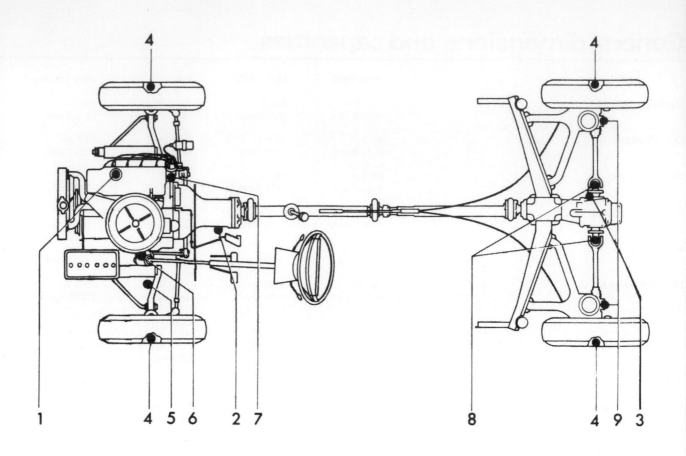

Recommended lubricants and fluids

Engine (1)	Castrol GTX
Gearbox (2):	
Manual	Castrol Hypoy Light (80 EP)
Automatic	Castrol TQ Dexron ®
Differential (3)	Castrol Hypoy B (90 EP)
Wheel bearings (4)	Castrol BNS Grease
Brake and clutch hydraulic systems (5)	Castrol Girling Universal Brake & Clutch Fluid
Steering box (6)	Castrol Hypoy B (90 EP)
Distributor (7)	Castrol GTX
Halfshaft UJ's (some models) (8)	Castrol LM Grease
Halfshaft sliding joints (some models) (9)	Castrol Hypoy B (90 EP)

Note: The above are general recommendations. Lubrication requirements vary from territory-to-territory and also depend on vehicle usage. Consult the operators handbook supplied with your car.

Chapter 1 Engine

Contents

Specifications

Engine type ... ... Four-in line, single overhead camshaft

Engine (general)

							1500	1502	1600	1602
Bore	...	...	...	...	...	...	3.228 in. (82.0 mm)	3.307 in. (84.0 mm)	3.307 in. (84.0 mm)	3.307 in. (84.0 mm)
Stroke	...	...	...	...	...	...	2.795 in. (71.0 mm)	2.795 in. (71.0 mm)	2.795 in. (71.0 mm)	2.795 in. (71.0 mm)
Capacity	...	...	...	...	...		91.47 cu. in (1499 cc)	95.99 cu. in (1573 cc)	95.99 cu. in (1573 cc)	95.99 cu. in (1573 cc)
Compression ratio	...	...	...	...			8.8 : 1	8.0 : 1	8.6 : 1	8.0 : 1
Max. output bhp (DIN)	...	...	...	...			80 @ 5700 rpm	75 @ 5800 rpm	83 @ 5500 rpm	85 @ 5700 rpm
Max. torque	...	...	...	...	...		86.8 lb/ft @ 3000 rpm	87 lb/ft @ 3700 rpm	91.2 lb/ft @ 3000 rpm	87 lb/ft @ 3700 rpm
							2000 touring	**2002**	**2002 TI**	**2002 TII**
Bore	...	...	...	...	...	...	3.504 in. (89.0 mm)	3.504 in. (89.0 mm)	3.504 in. (89.0 mm)	3.504 in. (89.0 mm)
Stroke	...	...	...	...	...	...	3.150 in. (80.0 mm)	3.150 in. (80.0 mm)	3.150 in. (80.0 mm)	3.150 in. (80.0 mm)
Capacity	...	...	...	...	...		121.44 cu. in (1990 cc)	121.44 cu. in (1990 cc)	121.44 cu. in (1990 cc)	121.44 cu. in (1990 cc)

	8.5 : 1	8.5 : 1	9.3 : 1	10 : 1
Compression ratio	8.5 : 1	8.5 : 1	9.3 : 1	10 : 1
Max. output bhp (DIN)	100 @ 5500 rpm	100 @ 5500 rpm	120 @ 5600 rpm	130 @ 5800 rpm
Max. torque	115.7 lb/ft @ 3000 rpm	115.7 lb/ft @ 3000 rpm	123 lb/ft @ 3000 rpm	130.2 lb/ft @ 4000 rpm

Cylinder block

Material		Cast-iron	

	1500	**1502, 1600 & 1602 series**	**2000 and 2002 series**
Bore:			
Standard	3.228 in. + 0.00087 in. (82.0 mm + 0.022 mm)	3.3071 in. + 0.00087 in. (84.0 mm + 0.022 mm)	3.5039 in. + 0.00087 in. (89.0 mm + 0.022 mm)
First rebore	+ 0.00984 in. (+ 0.25 mm)	+ 0.00984 in. (+ 0.25 mm)	+ 0.00984 in. (+ 0.25 mm)
Second rebore	+ 0.01969 in. (+ 0.50 mm)	+ 0.01969 in. (+ 0.50 mm)	+ 0.01969 in. (+ 0.50 mm)
Max. ovality	0.00039 in. (0.01 mm)	0.00039 in. (0.01 mm)	0.00039 in. (0.01 mm)
Max. taper	0.00039 in. (0.01 mm)	0.00039 in. (0.01 mm)	0.00039 in. (0.01 mm)

Pistons

Type	Aluminium domed crown	Aluminium domed crown	Aluminium recessed crown
Diameter (standard)	3.2258 in. (81.935 mm)	*Graded* A 3.3055 in. (83.96 mm) B 3.3059 in. (83.97 mm) C 3.3063 in. (83.98 mm)	*Graded* A 3.5024 in. (88.96 mm) B 3.5028 in. (88.97 mm) C 3.5032 in. (88.98 mm)
1st oversize	+ 0.0098 in. (+ 0.25 mm)	+ 0.0098 in. (+ 0.25 mm)	+ 0.0098 in. (+ 0.25 mm)
2nd oversize	0.01969 in. (0.50 mm)	0.01969 in. (0.50 mm)	0.01969 in. (0.50 mm)
Installed clearance:			
Mahle pistons	0.00256 in. (0.065 mm)	0.0016 in. (0.040 mm)	0.0016 in. (0.040 mm)
KS or Nural pistons	0.00236 in. (0.060 mm)	Not used	Not used

Piston rings

Top compression:			
End gap	0.0118 to 0.0177 in. (0.30 to 0.45 mm)	0.0118 to 0.0177 in. (0.30 to 0.45 mm)	0.0118 to 0.0177 in. (0.30 to 0.45 mm)
Side clearance	0.0024 to 0.0034 in. (0.060 to 0.087 mm)	0.00059 to 0.00114 in. (0.015 to 0.029 mm)	0.00059 to 0.00114 in. (0.015 to 0.029 mm)
Second compression:			
End gap	0.0118 to 0.0177 in. (0.30 to 0.45 mm)	0.0118 to 0.0177 in. (0.30 to 0.45 mm)	0.0118 to 0.0177 in. (0.30 to 0.45 mm)
Side clearance	0.0014 to 0.0024 in. (0.035 to 0.062 mm)	0.00047 to 0.00102 in. (0.012 to 0.026 mm)	0.00047 to 0.00102 in. (0.012 to 0.026 mm)
Third oil control:			
End gap	0.0098 to 0.0157 in. (0.25 to 0.40 mm)	0.0098 to 0.0157 in. (0.25 to 0.40 mm)	0.0098 to 0.0157 in. (0.25 to 0.40 mm)
Side clearance	0.00098 to 0.00205 in. (0.025 to 0.052 mm)	0.00043 to 0.00098 in. (0.011 to 0.025 mm)	0.00043 to 0.00098 in. (0.011 to 0.025 mm)

Gudgeon pins

Offset from piston centre-line	0.0591 in. (1.5 mm)	0.0591 in. (1.5 mm)	0.0591 in. (1.5 mm)
Diameter	0.8662 in. −0.00012 in. (22.0 mm −0.003 mm)	0.8662 in. −0.00012 in. (22.0 mm −0.003 mm)	0.8662 in. −0.00012 in. (22.0 mm −0.003 mm)
Clearance in piston	0.00012 to 0.00032 in. (0.003 to 0.008 mm)	0.00004 to 0.00020 in. (0.001 to 0.005 mm)	0.00004 to 0.00020 in. (0.001 to 0.005 mm)
Clearance in small end bush	0.00039 to 0.00071 in. (0.010 to 0.018 mm)	0.00012 to 0.00039 in. (0.003 to 0.010 mm)	0.00012 to 0.00039 in. (0.003 to 0.010 mm)

Crankshaft

Crankcase main bearing bore (red)	2.362 in. + 0.00039 in. (60.0 mm + 0.0010 mm)	2.362 in. +0.00039 in. (60.0 mm + 0.0010 mm)	2.362 in. + 0.00039 in. (60.0 mm + 0.0010 mm)

	1500	1502, 1600 & 1602 series	2000 and 2002 series
Crankcase main bearing bore (blue)	2.362 in. + 0.00039/0.00075 in. (60.0 mm + 0.010/0.019 mm)	2.362 in. + 0.00039/0.00075 in. (60.0 mm + 0.010/0.019 mm)	2.362 in. + 0.00039/0.00075 in. (60.0 mm + 0.010/0.019 mm)
Main bearing running clearance	0.0019 to 0.0027 in. (0.030 to 0.068 mm)	0.0019 to 0.0027 in. (0.030 to 0.068 mm)	0.0019 to 0.0027 in. (0.030 to 0.068 mm)
Main bearing journal diameter (standard):			
Red	2.165 in. −0.00098/0.00114 in. (55.0 mm −0.010/0.020 mm)	2.165 in. −0.00098/0.00114 in. (55.0 mm −0.010/0.020 mm)	2.165 in. −0.00098/0.00114 in. (55.0 mm −0.010/0.020 mm)
Blue	2.165 in. −0.00079/0.00114 in. (55.0 mm −0.020/0.029 mm)	2.165 in. −0.00079/0.00114 in. (55.0 mm −0.020/0.029 mm)	2.165 in. −0.00079/0.00114 in. (55.0 mm −0.020/0.029 mm)
Regrind stages (undersize):			
1	0.25 mm	0.25 mm	0.25 mm
2	0.50 mm	0.50 mm	0.50 mm
3	0.75 mm	0.75 mm	0.75 mm
Big-end bearing journal diameter	1.8898 in. − 0.00114/0.00098 in. (48.0 mm −0.009/0.025 mm)	1.8898 in. − 0.00114/0.00098 in. (48.0 mm −0.009/0.025 mm)	1.8898 in. − 0.00114/0.00098 in. (48.0 mm −0.009/0.025 mm)
Big-end bearing journals regrind stages (undersize):			
1	0.25 mm	0.25 mm	0.25 mm
2	0.50 mm	0.50 mm	0.50 mm
3	0.75 mm	0.75 mm	0.75 mm
Crankshaft endfloat	0.0024 to 0.0064 in. (0.06 to 0.18 mm)	0.00335 to 0.00685 in. (0.085 to 0.174 mm)	0.00335 to 0.00685 in. (0.085 to 0.174 mm)

Connecting rods

	1500	1502, 1600 & 1602 series	2000 and 2002 series
Overall length	5.315 in. (135.0 mm)	5.315 in. (135.0 mm)	5.315 in. (135.0 mm)
Big-end bearing running clearance	0.00114 to 0.00287 in. (0.029 to 0.073 mm)	0.00090 to 0.00272 in. (0.023 to 0.069 mm)	0.00090 to 0.00272 in. (0.023 to 0.069 mm)
Big-end bearing bore diameter	2.047 in. + 0.00039 in. (52.0 mm + 0.010 mm)	2.047 in. + 0.00039 in. (52.0 mm + 0.010 mm)	2.047 in. + 0.00039 in. (52.0 mm + 0.010 mm)

Camshaft

	1500	1502, 1600 & 1602 series	2000 and 2002 series
Bearing diameters	1.3780 in. (35.0 mm) 1.6536 in. (42.0 mm)	1.3780 in. (35.0 mm) 1.6536 in. (42.0 mm)	1.3780 in. (35.0 mm) 1.6536 in. (42.0 mm)
Running clearance	0.00134 to 0.00295 in. (0.034 to 0.075 mm)	0.00134 to 0.00295 in. (0.034 to 0.075 mm)	0.00134 to 0.00295 in. (0.034 to 0.075 mm)
Cam lift	0.2756 in. (7.0 mm)	0.2764 in. (7.02 mm)	0.2764 in. (7.02 mm)

Cylinder head

	1500	1502, 1600 & 1602 series	2000 and 2002 series
Material	Light alloy	Light alloy	Light alloy

Valves

	1500	1502, 1600 & 1602 series	2000 and 2002 series
Valve clearances - cold (inlet and exhaust) ...	0.006 to 0.008 in. (0.15 to 0.20 mm)	0.006 to 0.008 in. (0.15 to 0.20 mm)	0.006 to 0.008 in. (0.15 to 0.20 mm)
Valve overall length:			
Inlet	4.087 in $\pm$ 0.00079 (103.8 mm $\pm$ 0.2)	4.087 in $\pm$ 0.00079 (103.8 mm $\pm$ 0.2)	4.087 in $\pm$ 0.00079 (103.8 mm $\pm$ 0.2)
Exhaust	4.106 in $\pm$ 0.00079 (104.3 mm $\pm$ 0.2)	4.106 in $\pm$ 0.00079 (104.3 mm $\pm$ 0.2)	4.106 $\pm$ 0.00079 (104.3 mm $\pm$ 0.2)
Valve head diameter:			
Inlet	1.535 in. (39.0 mm)	1.654 in. (42.0 mm)	1.732 in.* (44.0 mm)*
Exhaust	1.378 in. (35.0 mm)	1.378 in. (35.0 mm)	1.496 in. (38.0 mm)

TII models 1.811 in. (46 mm)

	1500	1502, 1600 & 1602 series	2000 and 2002 series
Valve stem diameter (inlet and exhaust) ...	0.315 in. (8.0 mm)	0.315 in. (8.0 mm)	0.315 in. (8.0 mm)
Minimum edge thickness:			
Inlet	0.039 in $\pm$ 0.0004 (1.0 mm $\pm$ 0.1)	0.039 in $\pm$ 0.0004 1.0 mm $\pm$ 0.1)	0.039 in $\pm$ 0.0004 1.0 mm $\pm$ 0.1)
Exhaust	0.059 in. $\pm$ 0.0004 (1.5 mm $\pm$ 0.1)	0.059 in $\pm$ 0.0004 (1.5 mm $\pm$ 0.1)	0.059 in $\pm$ 0.0004 (1.5 mm $\pm$ 0.1)

	1500	1502, 1600 & 1602 series	2000 and 2002 series
Valve seat bore in cylinder head (dia):			
Inlet	1.654 in. (42.0 mm)	1.732 in. (44.0 mm)	1.85 in. (47.0 mm)
Exhaust	1.496 in. (38.0 mm)	1.496 in. (38.0 mm)	1.575 in. (40.0 mm)
External diameter seat:			
Inlet	1.660 in. (42.15 mm)	1.738 in. (44.15 mm)	1.856 in. (47.15 mm)
Exhaust	1.502 in. (38.15 mm)	1.502 in. (38.15 mm)	1.581 in. (40.15 mm)
Valve seat interference fit	0.00394 to 0.00591 in. (0.10 to 0.15 mm)	0.00394 to 0.00591 in. (0.10 to 0.15 mm)	0.00394 to 0.00591 in. (0.10 to 0.15 mm)
Valve seat angle	45°	45°	45°
Outer correction angle	15°	15°	15°
Inner correction angle	75°	75°	75°
Valve seat width:			
Inlet	0.063 to 0.079 in. (1.6 to 2.0 mm)	0.063 to 0.079 in. (1.6 to 2.0 mm)	0.063 to 0.079 in. (1.6 to 2.0 mm)
Exhaust	0.079 to 0.095 in. (2.0 to 2.4 mm)	0.079 to 0.095 in. (2.0 to 2.4 mm)	0.079 to 0.095 in. (2.0 to 2.4 mm)
Valve guide overall length	2.047 in. (52.0 mm)	2.047 in. (52.0 mm)	2.047 in. (52.0 mm)
Valve stem clearance in guide:			
Inlet	0.00098 to 0.00216 in. (0.025 to 0.055 mm)	0.00098 to 0.00216 in. (0.025 to 0.055 mm)	0.00098 to 0.00216 in. (0.025 to 0.055 mm)
Exhaust	0.00157 to 0.00275 in. (0.040 to 0.070 mm)	0.00157 to 0.00275 in. (0.040 to 0.070 mm)	0.00157 to 0.00275 in. (0.040 to 0.070 mm)
Wear limit	0.0059 in. (0.15 mm)	0.0059 in. (0.15 mm)	0.0059 in. (0.15 mm)
Guide external diameter (oversizes available)	0.5512 in. (14.0 mm)	0.5512 in. (14.0 mm)	0.5512 in. (14.0 mm)
Guide internal diameter	0.3150 in. (8.0 mm)	0.3150 in. (8.0 mm)	0.3150 in. (8.0 mm)
Guide projection in cylinder head	0.591 in. (15.0 mm)	0.591 in. (15.0 mm)	0.591 in. (15.0 mm)
Interference fit in head	0.00173 to 0.00059 in. (0.044 to 0.015 mm)	0.00272 to 0.00130 in. (0.069 to 0.033 mm)	0.00272 to 0.00130 in. (0.069 to 0.033 mm)

Valve springs

Wire thickness	0.167 in. (4.25 mm)	0.167 in. (4.25 mm)	0.167 in. (4.25 mm)
Coil external diameter	1.260 in. (31.9 mm)	1.260 in. (31.9 mm)	1.260 in. (31.9 mm)
Free-length	1.811 in. (46.0 mm)	1.7126 in. (43.5 mm)	1.7126 in. (43.5 mm)
Rocker shaft diameter	0.6103 in. (15.5 mm)	0.6103 in. (15.5 mm)	0.6103 in. (15.5 mm)
Rocker shaft running clearance	0.00063 to 0.00240 in. (0.016 to 0.061 mm)	0.00063 to 0.00303 in. (0.016 to 0.077 mm)	0.00063 to 0.00303 in. (0.016 to 0.077 mm)
Rocker arm running clearance	0.00063 to 0.00205 in. (0.016 to 0.052 mm)	0.00063 to 0.00205 in. (0.016 to 0.052 mm)	0.00063 to 0.00205 in. (0.016 to 0.052 mm)

Camshaft chain

Drive	Duplex roller chain 3/8 x 7/32 in.	Duplex roller chain 3/8 x 7/32 in.	Duplex roller chain 3/8 x 7/32 in.
Number of links.	94	94	94
Tensioner coil spring free-length	6.122 in. (155.5 mm)	6.122 in. (155.5 mm)	6.122 in. (155.5 mm)

Lubrication system

Type	Crankshaft driven gear or rotor type pump	Crankshaft driven gear or rotor type pump	Crankshaft driven gear or rotor type pump
Filter	Full-flow, external	Full-flow external	Full-flow external
Capacity (including filter)	7.54 Imp. pints; 4.25 litres; 9.13 US pints	7.54 Imp. pints; 4.25 litres; 9.13 US pints	7.54 Imp. pints; 4.25 litres; 9.13 US pints

Gear type oil pump

Pressure at idling	7 to 21 psi	7 to 21 psi	7 to 21 psi
Pressure at max. engine speed	71 to 85 psi	71 to 85 psi	71 to 85 psi
Gear tooth backlash:			
Normal	0.0012 to 0.0019 in. (0.03 to 0.05 mm)	0.0012 to 0.0019 in. (0.03 to 0.05 mm)	0.0012 to 0.0019 in. (0.03 to 0.05 mm)
Wear limit	0.0028 in (0.07 mm)	0.0028 in (0.07 mm)	0.0028 in (0.07 mm)
Gear endfloat	0.0019 in. (0.05 mm)	0.0019 in. (0.05 mm)	0.0019 in. (0.05 mm)
Wear limit	0.0028 in. (0.07 mm)	0.0028 in. (0.07 mm)	0.0028 in. (0.07 mm)
Pressure valve spring free-length	2.68 in. (68.0 mm)	2.68 in. (68.0 mm)	2.68 in. (68.0 mm)
Number of links in drive chain	44	44	44

Rotor type oil pump

Clearnace (outer rotor to housing) ...	0.0020 to 0.0079 in. (0.05 to 0.20 mm)	0.0020 to 0.0079 in. (0.05 to 0.20 mm)	0.0020 to 0.0079 in. (0.05 to 0.20 mm)

	1500	1502, 1600 & 1602 series	2000 and 2002 series
Rotor tip clearance	0.0035 to 0.0106 in. (0.09 to 0.27 mm)	0.0035 to 0.0106 in. (0.09 to 0.27 mm)	0.0035 to 0.0106 in. (0.09 to 0.27 mm)
Rotor to housing flange clearance	0.0013 to 0.0033 in. (0.034 to 0.084 mm)	0.0013 to 0.0033 in. (0.034 to 0.084 mm)	0.0013 to 0.0033 in. (0.034 to 0.084 mm)
Number of links in drive chain	46	46	46
Oil pressure:			
Idling	11 to 17 psi	11 to 17 psi	11 to 17 psi
At 4000 rpm	57 psi	57 psi	57 psi

Torque wrench settings

	lb/ft		Nm
Cylinder head nuts	54		75
Main bearing caps	45		62
Connecting rod big-end caps	40		55
Flywheel bolts	75		104
Chain tensioner plug	28		39
Crankshaft pulley nut	100		138
Distributor housing bolts (small)	8		11
Distributor housing bolts (large)	18		25
Rocker cover screws	8		11
Sump bolts	8		11
Timing cover bolts (small)	6		8
Timing cover bolts (large)	18		25
Bellhousing to engine bolts (small)	18		25
Bellhousing to engine bolts (large)	34		47
Engine mounting bracket to crankcase	34		47
Engine mounting centre bolts	18		25
Gearbox output flange nut	108		149
Propeller shaft front coupling nuts to output shaft flange	34		47

1 General description

The engine is of four cylinder, in-line, single overhead camshaft type. The combustion chambers are hemispherical.

The crankshaft has five main bearings and the lubrication system is based upon either a rotor or gear type oil pump which is chain driven from the crankshaft. A full-flow oil filter is incorporated in the system. The fuel system is either by carburettor or fuel injection according to model and cars destined for operation in North America incorporate an emission control system.

The engines of different capacities are all similar in construction but differ in bore and stroke as will be evident from the details given in the 'Specifications' Section.

The cylinder block is of cast-iron construction while the cylinder head is of light alloy. Valve seats and guides are replaceable, shrunk fit. The engine is inclined at 30° to lower the centre of gravity and to reduce the bonnet line.

2 Major operations possible with engine in position in car

1 The following components can be removed and refitted while the engine is still in the car. Where more than one major internal component is to be removed however, it will probably be quicker and easier to remove the engine complete.

Cylinder head and rocker shafts
Sump
Oil pump (after removal of sump)
Upper and lower timing gear covers
Timing cover oil seal
Crankshaft rear oil seal (after removal of gearbox and flywheel)
Piston/connecting rod assembly (after removal of cylinder head and sump)
Camshaft
Camshaft, chain and sprockets
Flywheel (or driveplate - automatic transmission) after gearbox and clutch removal
All ancillary components (alternator, distributor, water pump etc.)

3 Engine - removal method (general)

1 On carburettor models it is recommended that the engine is removed together with the gearbox (or automatic transmission), as a combined unit.

2 On fuel injection models it is recommended that the gearbox is first removed, as described in Chapter 6, and then the engine removed separately.

4 Engine (carburettor type) with manual gearbox - removal

1 Open the bonnet fully and mark the position of the hinge plates. With the help of an assistant, unbolt and remove the bonnet, storing it in a safe place where it will not slip and damage the paintwork.

2 Cover the top surfaces of the front wings with protective covers.

3 Drain the cooling system, retaining the coolant if it contains antifreeze and is required for further use.

4 Drain the engine oil and discard it.

5 Disconnect the leads from the battery terminals and remove the battery from the car. Always disconnect the earth lead first and reconnect last.

6 Identify the air cleaner hoses and their connections and remove the air cleaner according to type (Chapter 3), also the pre-heater from just to the rear of the radiator (photo).

7 Disconnect the radiator hoses from the thermostat and water pump and unbolt and remove the radiator from the engine compartment (photo).

8 Pull out the multi-pin plug from the back of the alternator and then disconnect the 'B+' cable from the alternator and the starter. On early 1500 models a dynamo was fitted (see Chapter 10) (photo).

9 Disconnect the lead from the water temperature transmitter switch (photo).

10 Disconnect the choke cable from the carburettor (photo).

11 *On models equipped with a Solex 32/32 DTDTA electrically pre-heated choke carburettor,* disconnect the lead.

12 *On cars equipped with an emission control system,* disconnect all hoses, leads and connections according to the system employed (Chapter 3).

13 Disconnect the fuel line from the inlet side of the fuel pump and plug the line.

14 Disconnect the heater hoses from the engine and tie them back out of the way.

15 Disconnect the lead from the oil pressure switch.

16 Disconnect the high tension (HT) leads from the spark plugs and the centre socket of the ignition coil.

17 Disconnect the low tension (LT) lead from the distributor.

18 Lift away the distributor cap complete with leads and then remove the rotor arm.

19 Disconnect the throttle linkage at the carburettor (photo).

20 *On 1500, 1600, 1602 models,* detach the pull rod from the clutch operating intermediate shaft then disconnect the intermediate shaft

bearing support from the transmission rear mounting crossmember. Withdraw the intermediate shaft and pushrod.

21 *On 2000 and 2002 models,* withdraw the clutch slave cylinder from the clutch bellhousing, release the hydraulic pipeline clip and tie the cylinder up out of the way. There is no need to disconnect the hydraulic circuit (photo).

22 Disconnect the speedometer drive cable from the gearbox, also the gearchange lever, as described in Chapter 6, according to type.

23 Disconnect the reversing lamp switch leads from the gearbox.

24 Disconnect the exhaust downpipe from the manifold, and release the pipe support and the front silencer box mountings (photo).

25 Disconnect the flexible drive coupling at the front of the propeller shaft from the gearbox rear driving flange. Hold the bolt heads quite still while unscrewing the self-locking nuts in order to avoid distorting the coupling. Release but do not remove, the propeller shaft centre bearing support screws, this will ease disconnection of the flexible coupling.

26 Unless the car is positioned over a pit, the front of the car must now be raised and securely supported to provide plenty of space to permit the gearbox to be lowered so that the engine/transmission can be hoisted from the engine compartment at a very steep angle.

27 Support the gearbox on a jack using a block of wood as an insulator.

28 Attach a hoist to the engine, using chains or slings, so positioned and of unequal length to ensure that the power unit will take on a steeply inclined attitude once the mountings are disconnected and the gearbox lowered.

29 Make sure that the weight of the engine is taken by the hoist and remove the two front mounting bolts (photos).

30 With the jack supporting the weight of the gearbox, remove the rear mounting crossmember from the gearbox and the bodyframe.

31 Remove the windscreen washer reservoir from the engine compartment and make a final check to ensure that all leads, controls and hoses have been disconnected.

32 Lower the gearbox jack in unison with raising the hoist and lift the engine/gearbox from the engine compartment. When the assembly is raised sufficiently, the car may be lowered and pushed backwards under the suspended engine/gearbox. The latter can then be lowered to the floor or placed on a suitable bench or stand (photo).

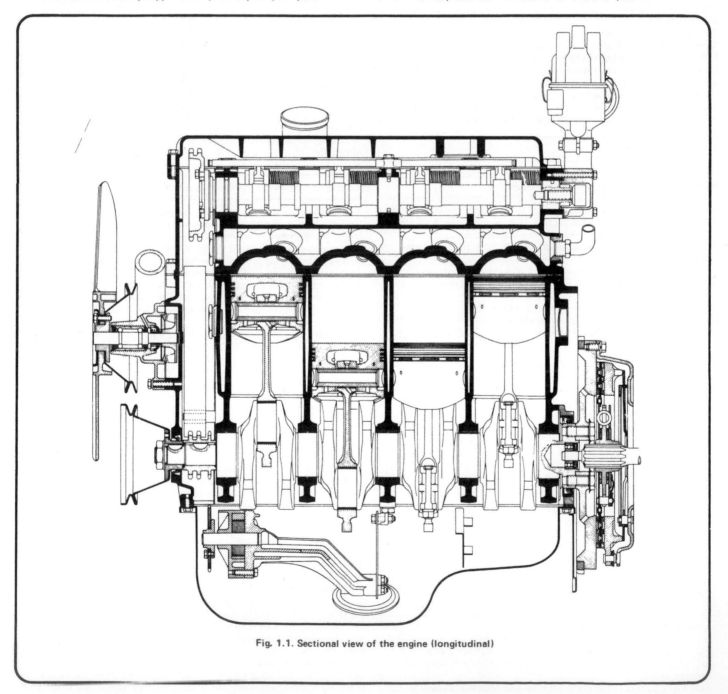

Fig. 1.1. Sectional view of the engine (longitudinal)

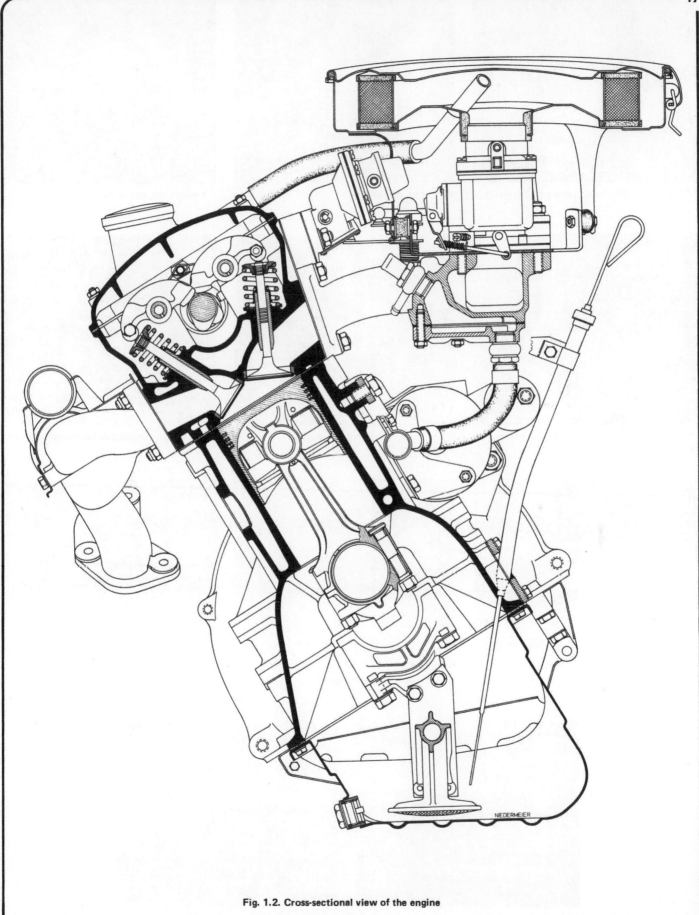

Fig. 1.2. Cross-sectional view of the engine

4.6 Removing air cleaner (single carburettor)

4.7 Removing radiator

4.8 Alternator terminals

4.9 Water housing with temperature sender switch

4.10 Choke cable connection (single carburettor)

4.19 Throttle connection (single carburettor)

4.21 Clutch slave cylinder (later 2000/2002 models)

4.24 Exhaust bracket to gearbox

4.29a Left-hand engine mounting

4.29b Right-hand engine mounting

4.32 Removing engine/gearbox

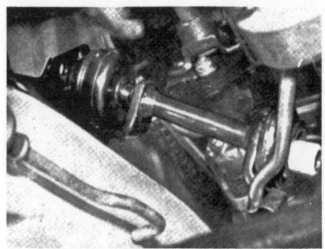

Fig. 1.3. Clutch operating components (1500/1600 series)

Fig. 1.5. Leads to starter inhibitor/reverse lamp switch (automatic transmission)

5 Engine (carburettor type) with automatic transmission - removal

1 The procedure is similar to that described in the preceding Section, but will require the following additional operations.
2 Drain the automatic transmission fluid and disconnect the transmission cooler lines (if fitted) at the radiator and plug the pipes.
3 Disconnect the lead from the automatic choke and the thermo-start valve.
4 Disconnect the leads from the reverse lamp/starter inhibitor switch and release the wiring from the clip on the transmission.
5 Disconnect the coolant hoses from the automatic choke housing.
6 Disconnect the downshift linkage by removing the clamp spring, return spring and cable retainer. Separate the linkage balljoint and pull the torsion shaft towards the front of the car.
7 Disconnect the speed selector control rod from the lower end of the selector lever by extracting the securing circlip.

6 Engine (fuel injection type) - removal

1 With this type of engine, due to the additional space required to clear the fuel injection components, it is recommended that the gearbox is removed from the car first, as described in Chapter 6.
2 Disconnect the leads from the battery and the alternator.
3 Remove the air filter.
4 Remove the radiator from the engine compartment.

Fig. 1.6. Throttle control and automatic choke hose connections at carburettor (automatic transmission)

1 Clamp spring   3 Cable retainer
2 Return spring

Fig. 1.4. Leads to automatic choke and thermo-start valve (automatic transmission)

Fig. 1.7. Disconnecting selector rod (automatic transmission)

5   Disconnect the fuel hose from the fuel injection pump and
disconnect the lead from the thermo-time switch.
6   Remove the main fuel filter from its front panel mounting.
Disconnect the fuel return hose.
7   Identify and mark the vacuum hose and disconnect it from the air
container.
8   Detach the lead from the start valve at the same time releasing the
cable from the clamps on the cam cover.
9   Disconnect the ignition high tension (HT) leads, pull out the
induction transmitter from the coil and then remove the distributor
cap and rotor arm.
10 Disconnect the accelerator linkage, also the starter motor lead.
11 Disconnect the heater hoses from the engine.
12 Remove the windscreen washer reservoir from the engine
compartment.
13 Fit chains or slings to the engine and take its weight on a hoist.
14 Remove both engine mounting centre stud nuts and release the
wiring harness clip from its location adjacent to the left-hand engine
mounting.
15 Lift the engine up and out of the engine compartment.

Fig. 1.10. Vacuum hose connection to air container and start valve
lead clamps (fuel injection)

Fig. 1.8. Air filter disconnection points (fuel injection)

Fig. 1.11. Induction transmitter at ignition coil (fuel injection)

Fig. 1.9. Hose connection to fuel pump and thermo-time switch lead
(fuel injection)

Fig. 1.12. Accelerator linkage and starter motor connections (fuel
injection)

7 Engine - separation from manual gearbox

1   Support the engine securely in the vertical position and then
unscrew and remove the bolts from the clutch bellhousing.
2   Loosen the transmission support bracket bolts and then extract the
bolts and cover which are located on the lower face of the bellhousing
(photo).
3   Pull the gearbox from the engine, supporting its weight and keeping
it square to the engine until the input shaft is clear of the clutch
mechanism, which is bolted to the rear face of the flywheel.

8 Engine - separation from automatic transmission

1   Loosen the support bracket bolts and remove the cover plate from
its location on the lower half of the front face of the torque converter
housing.
2   Unscrew each of the four driveplate to torque converter bolts.
These are only accessible one at a time and the crankshaft will have to
be turned to bring each bolt into view. To do this, apply a spanner to
the crankshaft pulley nut.
3   Support the weight of the automatic transmission unit and withdraw it to the rear, keeping it square to the engine. As the torque
converter housing is withdrawn, insert a piece of wood to act as a
lever in order to keep the torque converter from dropping out during
final withdrawal. Expect some loss of transmission fluid.

9 Dismantling - general

1   It is best to mount the engine on a dismantling stand but if one is
not available, then stand the engine on a strong bench so as to be at a
comfortable working height. Failing this, the engine can be stripped
down on the floor.
2   During the dismantling process the greatest care should be taken to
keep the exposed parts free from dirt. As an aid to achieving this, it is
a sound scheme to thoroughly clean down the outside of the engine,
removing all traces of oil and congealed dirt.
3   Use paraffin or a good water soluble solvent. The latter compound
will make the job much easier, as, after the solvent has been applied
and allowed to stand for a time, a vigorous jet of water will wash off
the solvent and all the grease and filth. If the dirt is thick and deeply
embedded, work the solvent into it with a wire brush.
4   Finally wipe down the exterior of the engine with a rag and only
then, when it is quite clean should the dismantling process begin. As
the engine is stripped, clean each part in a bath of paraffin or petrol.
5   Never immerse parts with oilways in paraffin (ie; the crankshaft),
but to clean, wipe down carefully with a petrol dampened rag. Oilways
can be cleaned out with wire. If an air line is present all parts can be
blown dry and the oilways blown through as an added precaution.
6   Re-use of old engine gaskets is false economy and can give rise to
oil and water leaks, if nothing worse. To avoid the possibility of
trouble after the engine has been reassembled **always** use new gaskets
throughout.
7   Do not throw the old gaskets away as it sometimes happens that an
immediate replacement cannot be found and the old gasket is then very
useful as a template. Hang up the old gaskets as they are removed on a
suitable hook or nail.
8   To strip the engine it is best to work from the top down. The sump
provides a firm base on which the engine can be supported in an
upright position. When this stage where the sump must be removed is
reached, the engine can be turned on its side and all other work carried
out with it in this position.
9   Wherever possible, replace nuts, bolts and washers fingertight from
wherever they were removed. This helps avoid later loss and muddle.
If they cannot be replaced then lay them out in such a fashion that it
is clear where they belong.

10 Engine ancillary components - removal

1   If a complete engine strip down is to be carried out, now is the
time to remove the following ancillary components from the unit. The

7.2 Transmission support bracket

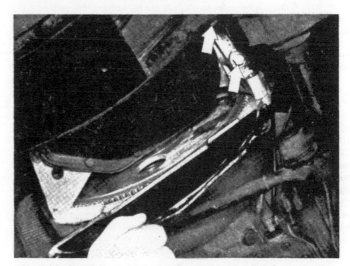
Fig. 1.13. Removing support bracket and cover plate from torque
converter housing (automatic transmission)

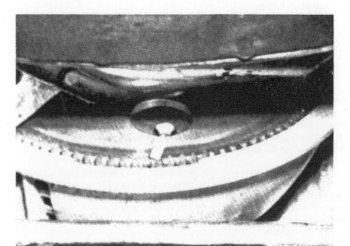
Fig. 1.14. Driveplate to torque converter bolt (automatic transmission)

removal operations are described in detail in the appropriate Chapters of this manual.

Generator and drivebelt (Chapter 10)
Starter motor (Chapter 10)
Inlet manifold (Chapter 3)
Exhaust manifold (Chapter 3)
Carburettor(s) (Chapter 3)
Water pump and fan (Chapter 2)
Fuel injection equipment (Chapter 3)
Fuel pump (Chapter 3)
Oil filter and dipstick (Section 35, of this Chapter)
Clutch mechanism (Chapter 5)
Emission control equipment, according to type (Chapter 3)

2   All the components can be removed from the engine while it is still in position in the car with the exception of the clutch assembly.

11 Cylinder head - removal and dismantling

If the engine is in the car, disconnect the battery, drain the cooling system, remove the air cleaner, disconnect the throttle and choke linkages. Disconnect the downfeed cable (automatic transmission) and electric choke leads where applicable. Disconnect emission control hoses. Disconnect the fuel inlet pipe at the fuel pump and plug the pipe. *On fuel injection models,* disconnect the fuel line at the injection pump. Disconnect the radiator hose from the thermostat housing and the hoses from the inlet manifold and branch pipe connection, also the heater hose from the cylinder head. Pull the leads from the water temperature sender and oil pressure switch. Remove the distributor cap complete with HT leads and on fuel injection models, the induction transmitter from the coil. Remove the oil dipstick support bracket. Disconnect the exhaust downpipe and support bracket from manifold and transmission respectively.

1   Remove the seven nuts which retain the rocker cover. Note the wiring harness clips under certain nuts. Lift off the cover (photos).
2   Remove the eight securing bolts and remove the timing gear upper cover. Note the attachment of the alternator earth wire under bolt no. 4.
3   Turn the crankshaft by applying a spanner to the crankshaft pulley nut until no. 1 piston is at TDC on its compression stroke. This can be checked by making sure that both valves on no. 1 cylinder are closed and that the marks on the distributor rotor arm and body rim are in alignment, also that the second (or left-hand when viewed from above) notch on the rim of the crankshaft pulley is opposite the pointer on the timing chain cover. Mark the relationship of the distributor body to the distributor drive housing, release the clamp bolt and withdraw the distributor.
4   Unscrew the chain tensioner plug and extract the spring and plunger.
5   Flatten the lockplates and remove the camshaft sprocket bolts at the same time maintaining tension on the chain in an upward direction so that it does not disengage from the teeth of the crankshaft sprocket. Retain the chain by wiring it to a convenient point on the crankcase.
6   Unscrew the ten cylinder head bolts, releasing them a turn at a time in diagonal sequence.
7   Lift the cylinder head complete with manifolds, carburettor or fuel injection components, as applicable. If the cylinder head is stuck tight, do not attempt to lever it off by inserting a tool in the gasket joint, but tap it gently all round with a soft-faced mallet or a heavy hammer using a block of hardwood interposed as an insulator.
8   To dismantle the cylinder head completely, remove the camshaft and rocker shafts, as described in Sections 12 and 13. To remove the valves from the cylinder head, compress the valve springs using a compressor with an extension which is necessary to reach the deeply recessed valve spring cap. Compress the valve spring and remove the split cotters then release the compressor gradually until the cap and valve spring can be extracted. Withdraw the valve from its guide (photos).
9   Remove all the valves in turn and retain them and their components in strict sequence for exact replacement. A sheet of stout card with holes punched in it and numbered 1 to 4 left (inlet) and 1 to 4 right (exhaust). No. 1 is at the front of the engine and left and right are when viewed from the driver's seat.
10 If full decarbonising is to be carried out, refer to Section 31.
11 If the inlet, exhaust manifolds or the fuel injection assemblies are to be removed from the cylinder head, refer to Chapter 3.

12 Camshaft - removal

If the engine is in the car, remove the cylinder head and withdraw the distributor, as described in Section 11. Remove the fuel pump and partially withdraw the operating rod.

If the engine is out of the car, the camshaft can be removed with the cylinder head still fitted to the block, provided that no piston is at TDC. Turn the crankshaft 90° from the no. 1 TDC position to achieve this. If this is not done, the valve heads may foul the pistons.
1   Remove the oil distribution tube (one bolt).
2   Release the rocker arm eccentric adjusters and open the valve clearances to the widest gap.
3   Before the camshaft can be withdrawn, all pressure on the cam lobes made by the rocker arms (which are depressing the open valves) must be released. To do this, a special tool is available from BMW dealers but an alternative method can be used. First make up two or three fork-ended tools from pieces of flat steel in accordance with the photograph. Turn the camshaft (using a lever between two bolts screwed into the end flange) until only three of the rocker arms are exerting pressure on the ends of their valve stems. With the help of an assistant, engage the fork-ended tools exactly as shown on the three rocker arms and depress them simultaneously. The camshaft can then be carefully withdrawn from its bearings and the thrust plate extracted. Do not over depress the rocker arms with the tools but only enough to release the rocker arm slides from their cam lobes (photos).
4   Unscrew and remove the two bolts which retain the camshaft guide plate and slide the plate downwards and withdraw it.

13 Rocker arms and shafts - removal

1   Remove the cylinder head and camshaft, as described in the preceding Sections.
2   Move the rocker arms and thrust washers to one side and extract the locating circlips.
3   With the distributor already removed, unbolt and withdraw the distributor drive housing complete with oil pressure switch from the rear face of the cylinder head. Note the location of the self-sealing washer and the special seal.
4   With a suitable drift, drive the rocker shafts out so that they emerge from the front of the cylinder head. Use a drift just less than the diameter of the rocker shaft, otherwise the shaft blanking plug will be driven into the shaft and block the oilways.
5   As each shaft is removed, extract the rocker arm components and keep them in strict order for exact refitting. Extracting sequence: spring - washer - rocker arm - collar - circlip.

14 Sump - removal

If the engine is in the car, drain the engine oil and with all 2002 models, remove the stabiliser bar. Attach a hoist to the engine and take its weight so that the engine mountings can be released. Turn the crankshaft so that no. 4 piston is at TDC.
1   Unscrew all the sump securing bolts in diagonally opposite sequence, a turn at a time and remove them.
2   If the sump is stuck tight, do not lever it off but cut round the gasket with a sharp knife.
3   Lift the sump from the crankcase. Where the engine is still in the car, draw the sump forward during removal in order to clear the oil pump intake pipe.

15 Oil pump - removal

1   Remove the sump, as described in the preceding Section.
2   Flatten the tabs of the lockwashers and unscrew and remove the oil pump sprocket bolts and then extract the sprocket from the loop of the driving chain.
3   Withdraw the oil pump after extracting the pump and pick-up tube securing bolts, the latter being screwed into no. 3 main bearing cap. Note the 'O' ring seal around the pressure relief pipe and the special seal at the pipe to crankcase flange (photos).

11.1a Location of spark plug lead clips on rocker cover

11.1b Location of spark plug lead clips on rocker cover

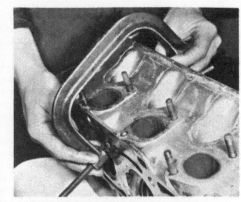
11.8a Removing a valve spring

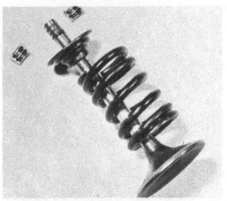
11.8b Valve components

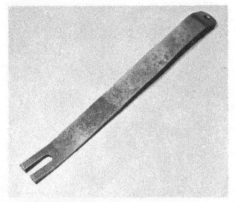
12.3a Tool for depressing rocker arms. Material is $\frac{1}{8}$ in (3 mm) steel plate

12.3b Fitting tool to rocker arm

12.3c Depressing a rocker arm

15.3a Oil pump

15.3b Oil pump pick-up tube bracket

15.3c Withdrawing oil pump from crankcase

15.3d Oil pump pressure relief pipe flange seal at crankcase

16.6 Alternator front mounting bracket

16 Timing components and covers - removal

1   The timing chain covers comprise an upper and lower section.
2   *To remove the upper cover,* first withdraw the rocker cover and then extract the eight bolts which secure the upper timing cover to the cylinder head. Note the location of the alternator earth lead under the centre right-hand cover bolt.
3   *To remove the lower cover,* disconnect the lead from the battery negative terminal, drain the cooling system and remove the water pump (Chapter 2).
4   Remove the upper timing cover, as described earlier in this Section.
5   Unscrew and remove the timing chain tensioner. To do this, unscrew the plug at the same time maintaining hand-pressure against the action of the internal coil spring, which will fly out if not restrained. Extract the piston and spring.
6   Disconnect the plug and lead from the rear of the alternator and then remove the alternator from its mountings, first having slipped the drivebelt from its pulleys. Unbolt the alternator mounting bracket.
7   Remove the cover plate from the lower half of the transmission bellhousing and jam the starter ring gear with a large screwdriver or cold chisel.
8   Unscrew and remove the crankshaft pulley nut and pull off the pulley (photo).
9   Unscrew and remove the lower timing cover bolts and withdraw the cover. Use a knife blade to separate the bottom edge of the cover from the sump gasket so that the gasket will not be broken or distorted. Note also the two locating dowels at the base of the cover and exercise care in removal.
10 If the main timing chain only, is to be removed, this can be carried out without removing the sump. Where the oil pump drive chain is also to be removed, then the sump will also have to be detached. Before removing either chain, mark the chain with a dab of paint or a piece of masking tape so that if it is going to be refitted it can be installed in the original running direction.
11 Remove the chain guide rails after extracting the retaining circlips. Remove the camshaft and oil pump sprockets to release the chains.
12 On some early models, a tensioning sprocket is used instead of a movable guide rail. The action of the tensioner plunger is identical in both types and the sprocket and tension lever can both be withdrawn after the circlips have been extracted.
13 The crankshaft sprocket can be removed after extracting the Woodruff key and the 'O' ring located behind it. The sprocket is a very tight fit and a suitable extractor will be required.

17 Piston/connecting rod - removal and dismantling

1   Remove the cylinder head and sump, as previously described in this Chapter.
2   Turn the crankshaft by applying a spanner to the pulley nut so that the big-end of the piston/connecting rod assembly in question is at the lowest point.
3   Each connecting rod and big-end cap is numbered at adjacent points on the same side (right-hand side facing front end of crankcase) commencing with no. 1 at the front of the engine.
4   Unscrew and remove the big-end nuts and then withdraw the cap and push the piston/connecting rod assembly out of the top of the block. If there is a severe wear ridge at the top of the cylinder bore, this should be scraped away before removing the piston, otherwise the piston rings may break.
5   Before separating the bearing shells from the cap and rod identify them in respect of each component using a piece of masking tape or a spirit marker on the backs of the shells. This is absolutely essential if the original shells are being refitted.
6   The gudgeon pin can be removed after extracting the circlips and pushing out the pin with finger pressure. Before doing this however, note the relationship of the rod oil hole (below small end) to the front facing arrow on the piston crown so that piston and rod can be reconnected the correct way round.
7   Discard the original big-end bolts and nuts and obtain new ones.
8   The piston rings are very brittle and will break easily if opened too far during removal. Two or three old feeler blades or strips of tin may be inserted behind each ring at equidistant points to facilitate removal. Use a twisting motion and pull the rings from the top of the piston.

16.8 Removing crankshaft pulley

Fig. 1.15. Timing chain tensioner sprocket (early models)
R Sprocket   S Circlip

The feeler blades will prevent a lower ring dropping into an empty groove as it is withdrawn.

18 Flywheel (or driveplate - automatic transmission) - removal

The flywheel or driveplate can be removed with the engine still in the car if the transmission unit is first withdrawn (see Chapter 6).
1   Remove the clutch assembly (manual gearbox).
2   Mark the relative position of the flywheel or driveplate to the crankshaft rear flange.
3   Lock the starter ring gear with a heavy screwdriver or cold chisel and unscrew and remove the securing bolts. Some flywheels have eight securing bolts, these must be discarded and new ones obtained. On six bolt types, the lockplate must be renewed.

19 Crankshaft and main bearings - removal

1   With the engine removed, withdraw the clutch assembly, flywheel, timing chain and oil pump.
2   Unbolt and remove the rear oil seal retainer.
3   Normally the crankshaft will only be removed at the time of major engine overhaul when the piston/connecting rods will already have been removed but if required, the crankshaft and main bearings can be withdrawn without disturbing the cylinder head or piston/connecting rod assemblies, providing the big-ends are disconnected and the pistons pushed a little way up the bores, **from their lowest position.** Do not push the pistons too far up the bores, or the rings will be ejected from the bores and full dismantling will then be necessary.

4   The main bearing caps are numbered from 1 to 5, starting at the front of the engine and the numerals being legible from the same direction. Do not confuse these numbers with the casting marks.

5   Unscrew and remove the main bearing bolts and lift off the caps complete with bearing shells. Identify the shells in respect of position. Note the oil pump pick-up pipe support attached to no. 3 main bearing. The shells in this bearing are flanged and act as the crankshaft thrust washers.

6   Lift the crankshaft from the crankcase, taking care that the bearing shells are retrieved and identified with regard to location if they are to be refitted.

20 Examination and renovation - general

With the engine stripped down and all parts thoroughly cleaned, it is now time to examine everything for wear. The items should be checked and where necessary renewed or renovated as described in the following Sections.

21 Crankshaft and main bearings - examination and renovation

1   Examine the crankpin and main journal surfaces for signs of scoring or scratches. Check the ovality of the crankpins at different positions with a micrometer. If more than specified out of round, the crankpin will have to be reground. It will also have to be reground if there are any scores or scratches present. Also check the journals in the same fashion.

2   If it is necessary to regrind the crankshaft and fit new bearings your local BMW garage or engineering works will be able to decide how much metal to grind off and the size of new bearing shells. It is also necessary, at least in theory, for the surface hardening (Nitriding) of the bearing surfaces to be restored.

3   Full details of crankshaft regrinding tolerances and bearing under-sizes are given in Specifications.

4   The main bearing clearances may be established by using a strip of 'Plastigage' between the crankshaft journals and the main bearing/shell caps. Tighten the bearing cap bolts to a torque of 45 lb/ft (62 Nm). Remove the cap and compare the flattened Plastigage strip with the index provided. The clearance should be compared with the tolerances in Specifications.

5   Temporarily refit the crankshaft to the crankcase having refitted the upper halves of the shell main bearings in their locations. Fit no. 3 main bearing cap only, complete with shell bearing and tighten the securing bolts to 45 lb/ft (62 Nm) torque. Using a feeler gauge, check the endfloat by pushing and pulling the crankshaft. Where the endfloat is outside the specified tolerance, no. 3 bearing shells will have to be renewed.

6   Finally check the pilot ball race in the centre of the crankshaft rear flange. If it is worn or damaged, extract it.

7   When installing the new pilot bearing components, pack the bearing with high-melting point grease and impregnate the felt ring with hot tallow. Make sure that the marking on the cover plate faces outwards and then tap the retainer (4) right home until it seats.

22 Connecting rods and bearings - examination and renovation

1   Big-end bearing failure is indicated by a knocking from within the crankcase and a slight drop in oil pressure.

2   Examine the big-end bearing surfaces for pitting and scoring. Renew the shells in accordance with the sizes specified in Specifications. Where the crankshaft has been reground, the correct undersize big-end shell bearings will be supplied by the repairer.

3   Should there be any suspicion that a connecting rod is bent or twisted, it must be replaced by one of similar weight. Without bearing shells, the new rod must be within 0.14 oz (4 g) of the weight of the original component.

4   The connecting rod small end bush can be pressed out and a new one installed. Make sure that the seam in the bush is at 90° to the small oil hole to provide correct alignment of the oil drillings. No reaming is required after installation.

5   Measurement of the big-end bearing clearances may be carried out in a similar manner to that described for the main bearings in the

Fig. 1.16. Crankshaft rear spigot bearing components

| 1 | Ball bearing | 3 | Felt ring |
| 2 | Cover plate | 4 | Retainer |

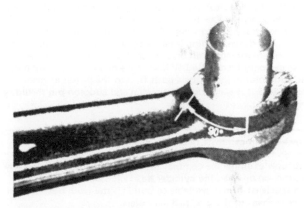

Fig. 1.17. Correct positioning of small end bush before pressing in

previous Section, but tighten the securing nuts on the cap bolts to 38 lb/ft (53 Nm).

23 Cylinder bores - examination and renovation

1   The cylinder bores must be examined for taper, ovality, scoring and scratches. Start by carefully examining the top of the cylinder bores. If they are at all worn a very slight ridge will be found on the thrust side. This marks the top of the piston ring travel. The owner will have a good indication of the bore wear prior to dismantling the engine, or removing the cylinder head. Excessive oil consumption accompanied by blue smoke from the exhaust is a sure sign of worn cylinder bores and piston rings.

2   Measure the bore diameter just under the ridge with a micrometer and compare it with the diameter at the bottom of the bore, which is not subject to wear. If the difference between the two measurements is more than 0.008 in (0.2032 mm) then it will be necessary to fit special pistons and rings or to have the cylinders rebored and fit oversize pistons. If no micrometer is available remove the rings from a piston and place the piston in each bore in turn about ¾ in below the top of the bore. If an 0.0012 in (0.0254 mm) feeler gauge slid between the piston and the cylinder wall requires more than a pull of between 1.1 and 3.3 lbs (0.5 and 1.5 kg) to withdraw it, using a spring balance then remedial action must be taken. Oversize pistons are available as listed in Specifications.

3 These are accurately machined to just below the indicated measurements so as to provide correct running clearances in bores bored out to the exact oversize dimensions.
4 If the bores are slightly worn but not so badly worn as to justify reboring them, then special oil control rings and pistons can be fitted which will restore compression and stop the engine burning oil. Several different types are available and the manufacturer's instructions concerning their fitting must be followed closely.
5 If new pistons are being fitted and the bores have not been reground, it is essential to slightly roughen the hard glaze on the sides of the bores with fine glass paper so the new piston rings will have a chance to bed in properly.

24 Pistons and piston rings - examination and renovation

1 If the original pistons are to be refitted, carefully remove the piston rings as described in Section 1.
2 Clean the grooves and rings free from carbon, taking care not to scratch the aluminium surfaces of the pistons.
3 If new rings are to be fitted, then order the top compression ring to be stepped to prevent it impinging on the 'wear ring' which will almost certainly have been formed at the top of the cylinder bore.
4 Before fitting the rings to the pistons, push each ring in turn down to the part of its respective cylinder bore (use an inverted piston to do this and to keep the ring square in the bore) and measure the ring end gap. The gaps should be as listed in Specifications Section.
5 Now test the side clearance of the compression rings which again should be as shown in Specifications Section.
6 Where necessary a piston ring which is slightly tight in its groove may be rubbed down, holding it perfectly squarely on an oilstone or sheet of fine emery cloth laid on a sheet of plate glass. Excessive tightness can only be rectified by having the grooves machined out.
7 The gudgeon pin should be a push fit into the piston at room temperature. If it is slack, both the piston and gudgeon pin should be renewed.

25 Camshaft and camshaft bearings - examination and renovation

1 Carefuoly examine the camshaft bearings for wear. If there is any pitting, scoring or wear, the cylinder head will have to be renewed unless a specialist firm is available to build up the worn bearings and in-line bore them to the specified diameters.
2 The camshaft itself should show no signs of wear, but, if very slight scoring on the cams is noticed, the score marks can be removed by very gentle rubbing down with a very fine emery cloth. The greatest care should be taken to keep the cam profiles smooth.
3 Examine the skew gear for wear, chipped teeth or other damage.

26 Valves and valve seats - examination and renovation

1 Examine the heads of the valves for pitting and burning, especially the heads of the exhaust valves. The valve seatings should be examined at the same time. If the pitting on valve and seat is very slight the marks can be removed by grinding the seats and valves together with coarse, and then fine, valve grinding paste.
2 Where bad pitting has occurred to the valve seats it will be necessary to recut them and fit new valves. If the valve seats are so worn that they cannot be recut, then it will be necessary to fit new valve seat inserts. These latter two jobs should be entrusted to the local BMW agent or engineering works. In practice it is very seldom that the seats are so badly worn that they require renewal. Normally, it is the valve that is too badly worn for replacement, and the owner can easily purchase a new set of valves and match them to the seats by valve grinding.
3 Valve grinding is carried out as follows: Smear a trace of coarse carborundum paste on the seat face and apply a suction grinder tool to the valve head. With a semi-rotary motion, grind the valve head to its seat, lifting the valve occasionally to redistribute the grinding paste. When a dull matt even surface finish is produced on both the valve seat and the valve, wipe off the paste and repeat the process with fine carborundum paste, lifting and turning the valve to redistribute the paste as before. A light spring placed under the valve head will greatly

ease this operation. When a smooth unbroken ring of light grey matt finish is produced, on both valve and valve seat faces, the grinding operation is completed.
4 Scrape away all carbon from the valve head and the valve stem. Carefully clean away every trace of grinding compound, taking great care to leave none in the ports or in the valve guides. Clean the valves and valve seats with a paraffin soaked rag then with a clean rag, and finally if an air line is available, blow the valves, valve guides and valve ports clean.

27 Valve guides - examination and renovation

1 Test each valve in its guide for wear. After a considerable mileage, the valve guide bore may wear oval. This can best be tested by inserting a new valve in the guide and moving it from side to side. If the tip of the valve stem deflects by about 0.0080 in (0.2032 mm) then it must be assumed that the tolerance between the stem and guide is greater than the permitted maximum as listed in Specification Section.
2 New valve guides are available in diameters of oversize as specified.
3 To remove and install valve guides, the cylinder head should be heated to between 220 and 248ºC (104 and 120ºC) in a domestic oven. Drive the old guides out towards the combustion chamber and then ream the valve guide holes in the cylinder head to specification.
4 Press in the new guides from the camshaft side. The valve guides must project 0.5906 in (15.0 mm) above the surface of the cylinder head on the camshaft side when installed, and then ream them to specification.
5 Unless the necessary reamers are available, it is preferable to leave valve guide renewal to your BMW dealer.

28 Timing chain and gears - examination and renovation

1 Examine the teeth on both the crankshaft gear wheel and the camshaft gear wheel for wear. Each tooth forms an inverted 'V' with the gearwheel periphery, and if worn the side of each tooth under tension will be slightly concave in shape when compared with the other side of the tooth (ie; one side of the inverted 'V' will be concave when compared with the other). If any sign of wear is present the gearwheels must be renewed.
2 Examine the links of the chain for side slackness and renew the chain, if any slackness is noticeable when compared with a new chain. It is a sensible precaution to renew the chain at about 30,000 miles (48,000 km) and at a lesser mileage if the engine is stripped down for a major overhaul. The actual rollers on a very badly worn chain may be slightly grooved.

29 Rocker arms and rocker shafts - examination and renovation

1 Thoroughly clean the rocker shaft and then check the shaft for straightness by rolling it on plate glass. It is most unlikely that it will deviate from normal, but if it does, purchase new shafts. The surface of the shaft should be free from any worn ridges caused by the rocker arms. If any wear is present, renew the shafts.
2 Check the rocker arms for wear of the rocker bushes, for wear at the rocker arm face which bears on the valve stem, and for wear of the slide faces. Wear in the rocker arm bush can be checked by gripping the rocker arm tip and holding the rocker arm in place on the shaft, noting if there is any lateral rocker arm shake. If shake is present, and the arm is very loose on the shaft, a new bush or rocker arm must be fitted.
3 Check the roller which bears on the end of the valve stem, also the slide pads which bear on the cam lobes and renew as necessary.

30 Flywheel (or driveplate - automatic transmission) - starter ring gear - examination and renovation

1 If the teeth of the starter ring gear are badly worn, the ring gear will have to be renewed on a flywheel or the complete driveplate renewed on an automatic transmission unit.
2 To remove the ring gear from a flywheel, drill a hole (6.0 mm in diameter) at the root of one tooth. Do not drill right through (8.0 mm in depth should be enough).

3 Split the ring with a sharp cold chisel.
4 Heat the new ring gear in an oil bath or electric oven to between 400 and 450°F (200 and 230°C).
5 Place the ring on the flywheel (chamfer towards engine) and tap it squarely into position using a brass drift.
6 Where the machined face of the flywheel is scored or shows surface cracks, it should be surface ground but the thickness of the driven plate contact area of the flywheel must never be reduced below 0.535 in (13.6 mm).

---

31 Cylinder head - decarbonising and examination

1 With the cylinder head removed, use a blunt scraper to remove all trace of carbon and deposits from the combustion spaces and ports. Remember that the cylinder head is aluminium alloy and can be damaged easily during the decarbonising operations. Scrape the cylinder head free from scale or old pieces of gasket or jointing compound. Clean the cylinder head by washing it in paraffin and take particular care to pull a piece of rag through the ports and cylinder head bolt holes. Any dirt remaining in these recesses may well drop onto the gasket or cylinder block mating surface as the cylinder head is lowered into position and could lead to a gasket leak after reassembly is complete.
2 With the cylinder head clean, test for distortion if a history of coolant leakage has been apparent. Carry out this test using a straight edge and feeler gauges or a piece of plate glass. If the surface shows any warping in excess of 0.039 in (0.1015 mm) then the cylinder head will have to be resurfaced which is a job for a specialist engineering company. The depth of the cylinder head must not be reduced by more than 0.0197 in (0.5 mm) and the depth of the upper timing cover must be reduced by an equivalent amount to compensate.
3 Clean the pistons and top of the cylinder bores. If the pistons are still in the block then it is essential that great care is taken to ensure that no carbon gets into the cylinder bores as this could scratch the cylinder walls or cause damage to the piston and rings. To ensure this does not happen, first turn the crankshaft so that two of the pistons are at the top of their bores. Stuff rag into the other two bores or seal them off with paper and masking tape to prevent particles of carbon entering the cooling system and damaging the water pump.
4 Rotate the crankshaft and repeat the carbon removal operations on the remaining two pistons and cylinder bores.
5 Thoroughly clean all particles of carbon from the bores and then inject a little light oil round the edges of the pistons to lubricate the piston rings.

---

32 Oil pump (gear type) - examination and renovation

1 Early models were fitted with a gear type oil pump.
2 Dismantling is similar to the description for rotor type pumps given in the next Section.
3 The backlash of the gear teeth should not exceed 0.003 in (0.07 mm) and the faces of the gears should not be below the flange of the housing by an amount exceeding 0.002 in (0.05 mm). Use feeler blades and a straight edge to check this.

---

33 Oil pump - changing from gear type to rotor type

1 Should the components of a gear type pump become badly worn and spares cannot be obtained, a new rotor type pump can be installed provided the following modifications are made.
2 Refer to Fig. 1.21. Having removed the old oil pump, extract the pivot pin (1), washer (2) and plug (3).
3 Insert the sealing washer (4) and secure the pipe with a new pivot pin (6) and lockwasher (7). These components should be obtained at the same time that the new pump is purchased.
4 Install the new rotor type pump, sprocket and 24 link chain. If necessary insert a shim (9) noting carefully the hole for alignment with the oil hole, so that the driving chain can be slightly deflected when light thumb pressure is applied.

Fig. 1.18. Drilling hole in starter ring gear

Fig. 1.19. Splitting starter ring gear

---

34 Oil pump (rotor type) - examination and renovation

1 Unbolt and remove the cover from the oil pump (photo).
2 Unscrew and remove the plug from the oil pressure relief valve and extract the plunger and spring (photo).
3 Using feeler blades check the following clearances:
   a) *Between the outer rotor and pump housing which should be from 0.0039 to 0.0059 in (0.1 to 0.15 mm).*
   b) *Between the tips of the inner and outer rotors which should be from 0.0047 to 0.0079 in (0.12 to 0.20 mm).*
   c) *Between the rotor face and the cover mating face of the pump housing which should be from 0.0014 to 0.0043 in (0.036 to 0.109 mm).*
   Where these tolerances are exceeded, renew the components as necessary (photos).
4 If the drive flange must be removed, use a two legged extractor. Press on the new flange so that the distance between the outer faces of flange and rotor is 1.681 in (42.7 mm).

---

35 Oil filter - renewal

1 The oil filter may be one of three different types:
A With gear type oil pumps, a bolt-on type filter bowl and separate

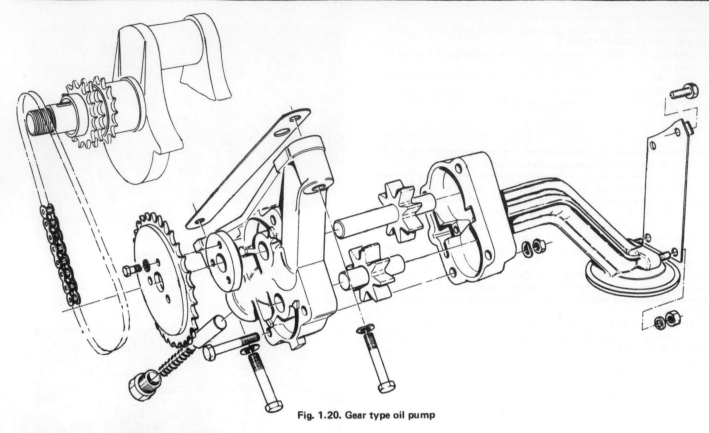

Fig. 1.20. Gear type oil pump

Fig. 1.21. Identification of components affected when changing from
gear type to rotor type oil pump

1	Old pivot pin	5	Pipe
2	Washer	6	New pivot pin
3	Plug	7	Lock washer
4	New sealing washer		

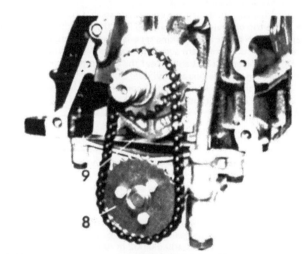

Fig. 1.22. Location of oil pump chain tensioning shim (9) and
sprocket (8)

filter element is used. This is a Mann and Hummel unit and can be
identified by the 7.0 mm diameter oil passage in the upper housing.
The pressure relief valve can be dismantled.

B With rotor type oil pumps, a bolt-on type filter of Purolator
manufacture may be encountered (this is identified from the Mann and
Hummel version by its 12.0 mm diameter oil passage in the upper
housing). The pressure relief valve in this type of filter cannot be
dismantled.

C On later models with rotor type oil pumps, a disposable type screw-
on filter cartridge is used in conjunction with a light alloy housing
(photos).

36 Oil seals - renewal

1   At the time of major engine overhaul, always renew the crankshaft
front and rear oil seals.

2   The front seal which is located in the timing chain lower cover can
be extracted using a hooked tool after the crankshaft pulley has been
removed or if the timing cover has been removed the seal can be
removed with a piece of tubing used as a drift (photo).

3   The rear seal is housed in a retainer accessible after the clutch and
flywheel (or driveplate - automatic transmission) have been removed
(photo).

34.1 Dismantling oil pump

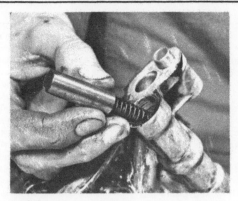

34.2 Oil pressure relief valve plunger and spring

34.3a Testing oil pump outer rotor to housing wear

34.3b Testing inner to outer lobe wear

34.3c Testing oil pump rotor endfloat

35.1a Cartridge type oil filter

35.1b Removing oil filter housing

36.2 Timing lower cover oil seal

36.3 Installing crankshaft rear bearing oil seal to retainer

37 Crankcase ventilation system

1   A positive type system is installed on all but the earliest models. The system ensures that the blow-by gases which accumulate in the engine crankcase are drawn out through the rocker cover and into the air cleaner where they are sucked into the engine combustion chambers and burned during the normal firing stroke.

2   Various versions of the system are used according to engine type but there is no maintenance, except to occasionally check the security of the connecting hoses (see also Chapter 3, Section 29).

38 Engine - preparation for reassembly

1   To ensure maximum life with reliability from a rebuilt engine, not only must everything be correctly assembled but all components must be spotlessly clean and the correct spring or plain washers used where originally located. Always lubricate bearing and working surfaces with clean engine oil during reassembly of engine parts.

2   Before reassembly commences, renew any bolts or studs, the threads of which are damaged or corroded.

3   As well as your normal tool kit, gather together clean rags, oil can, a torque wrench and a complete (overhaul) set of gaskets and oil seals.

39 Crankshaft and main bearings - installation

1 With the crankcase inverted, insert the bearing shells into their crankcase recesses. Make sure that both sides of each shell is absolutely clean before installing it. No. 3 shells incorporate the thrust flanges (photo).

2 Oil the bearings liberally and carefully lower the crankshaft into position (photos).

3 Fit the bearing shells to the main bearing caps and install them in their correct sequence (numbered and legible from the front of the engine) (photo).

4 Screw in and tighten the main bearing cap bolts to the specified torque (photo).

5 Check that the crankshaft turns smoothly with hand-pressure and then install the rear oil seal retainer complete with new oil seal and gasket (photo).

40 Flywheel (or driveplate - automatic transmission) - installation

1 Install the flywheel to the crankshaft flange so that marks made before removal are in alignment.

2 Where eight securing bolts are used, the old bolts must be discarded and new ones used, smearing them with thread sealing compound (photo).

3 Where six securing bolts are used, use a new locking plate.

4 Once the bolts are screwed in finger-tight, jam the starter ring gear and tighten the bolts to the specified torque wrench settings (photo).

41 Piston/connecting rods - installation

1 Assemble no. 1 piston to the connecting rod, making sure that with the arrow on the piston crown facing forwards (as if in the engine) the oil hole just below the small end bush on the rod faces in the same direction as the arrow. The sequence numbers on cap and rod should face to the right (looking to the front of the engine) (photo).

2 Push the gudgeon pin into position using finger-pressure and insert two new circlips (photo).

3 Install the piston rings by reversing the removal process. Note the sequence of fitting and ring cross sections from the diagram.

4 Stagger the piston ring end gaps at equal points of a circle.

5 Insert the bearing shell into the connecting rod, apply oil liberally to the bearing and the piston rings and smear some up and down the cylinder bore and fit a piston ring compressor (photo).

6 Insert the connecting rod into the cylinder bore taking care not to scratch the bore surfaces. With the compressor standing squarely on the top of the cylinder block and the piston rings well compressed (but not tight), place the wooden shaft of a hammer on the piston crown and then give the head of the hammer a sharp tap to drive the piston/rod assembly down the bore. Remove the ring compressor (photo).

7 Turn the crankshaft so that the number 1 big-end journal is at its lowest point of travel and then pull down the connecting rod onto it. Make sure that the bearing shell is not displaced.

8 Fit the shell to the big-end cap, oil it and install it, making sure that the numbers on cap and rod are adjacent. Recheck that the piston crown arrow faces the front of the engine (photos).

9 Install the big-end bolts and nuts and tighten to specified torque. The narrower diameter of the nut is nearest the big-end cap (photo).

10 Repeat the operations on the three remaining piston/connecting rod assemblies.

39.1 Fitting no. 3 main bearing shell to crankcase

39.2a Lubricating main bearing shells in crankcase

39.2b Installing crankshaft

39.3 Installing no. 3 main bearing cap. Numbers visible here are casting marks

39.4 Tightening a main bearing cap bolt

39.5 Installing crankshaft rear bearing oil seal retainer

40.2 Flywheel installed

40.4 Tightening a flywheel bolt

41.1a Directional mark on piston crown

41.1b Oil hole in connecting rod

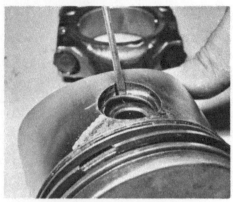

41.2 Installing a gudgeon pin circlip

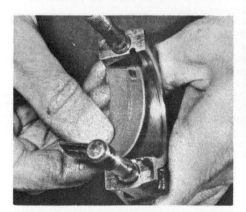

41.5 Fitting a bearing shell to a connecting rod

41.6 Installing a piston fitted with a piston ring compressor

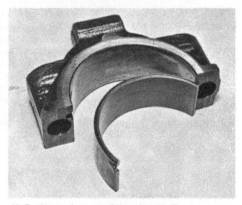

41.8a Big-end cap and bearing shell

41.8b Fitting a big-end cap and shell

41.8c Big-end identification numbers

41.9 Tightening big-end nuts

42 Timing components, covers and oil pump - installation

1 If the crankshaft sprocket has been removed, tap or press it into position (photo).
2 Fit a new 'O' ring seal to its front face and insert the Woodruff key into the crankshaft keyway.
3 Fit the chain guide rails and retaining circlips (photos).
4 If the oil pump and drive chain have been removed, refit the oil pump drive chain to the rear teeth of the crankshaft sprocket and then install the double roller timing chain to the front teeth of the sprocket. Let the oil pump chain hang downwards pending installation of the oil pump and draw the timing chain upwards and hang it over the top of the cylinder block. If the original chains are being refitted, make sure that the marks made before removal to indicate their running direction are correctly positioned.
5 If the oil pump was removed, now is the time to refit it. The operations are a reversal of removal but make sure that the 'O' ring seal round the pressure relief pipe is in position.
6 Locate the oil pump chain sprocket within the loop of the drive chain and bolt it to the pump flange using new lockwashers. At this stage check the chain tension by applying light thumb pressure. Only a slight deflection should be observed. If the chain is slack a shim should be installed but note the location of the shim oil hole (photo).
7 Install the timing lower cover complete with new oil seal (Section 36) and having applied some jointing compound to the new joint gaskets. If the cover was removed without disturbing the sump, apply jointing compound to the exposed part of the sump gasket. If the front end of the sump gasket was damaged during removal of the cover, a section can be cut from a new sump gasket and it can be fixed in position with jointing compound, this will save removing the sump if the engine is in the car (photos).
8 Install the crankshaft pulley and nut, tightening it to torque after jamming the starter ring gear (photos).
9 Install the timing chain. If the cylinder head has been removed, install it together with the camshaft and rocker shafts (Section 44 and 45). Insert the camshaft sprocket within the loop of the timing chain and fit the sprocket to the camshaft flange. The following conditions must apply. No. 1 piston must be at TDC (pointer opposite second notch of crankshaft pulley) and the camshaft flange dowel pin hole must be at its lowest point. This is indicated by a mark at the top of the flange being in alignment with a mark cast into the cylinder head (photos).
10 A certain amount of repositioning of the camshaft sprocket within the loop of the timing chain will be necessary until the sprocket will engage with the camshaft flange without having to move the flange in either direction. On no account move the crankshaft from its TDC position.
11 Install the camshaft bolts using new locking plates, tighten to specified torque and bend up the tabs of the locking plates.
12 Install the chain tensioner plunger, spring and plug (photos).
13 Install the timing upper cover, having smeared its mating flanges with jointing compound. Install the bolts which connect with the lower cover only finger-tight until the screws which retain the upper cover to the cylinder head have been tightened to the specified torque.

43 Sump - installation

1 Stick the new sump gasket into position using jointing compound. Apply it liberally to the joints and corners of the timing lower cover mating flange (photo).
2 Smear the sump mating flanges with jointing compound and offer it into position. Screw in all the bolts finger-tight and then tighten them to the specified torque a turn at a time and in diagonally opposite sequence.

44 Cylinder head - reassembly of valves, rocker shafts and camshaft

1 Install the first valve to its guide having applied engine oil to the valve stem.
2 If the engine has covered 25000 miles (40000 km) on the original valve springs, renew them, also the valve stem oil seals on both inlet and exhaust valves.

3 Install a spring lower cup, a valve stem oil seal, a spring, the cap and after compressing the spring insert the split cotters (photos).
4 Remove the compressor and then repeat the operations on the remaining seven valves making sure that they are all returned to their original (or ground-in locations - if new valves are being installed).
5 Drive the rocker shafts into their original locations, fitting the rocker arms and springs also in their original positions as the shaft passes through (photo).
6 Make sure that the rocker shafts are aligned to receive the cylinder head bolts into the shaft cut-outs. The rear end of the inlet rocker shaft is open but the rear end of the exhaust rocker shaft must be closed. A plug is used for this. Fit the rocker shaft circlips (photos).
7 Install the distributor drive housing to the rear end of the cylinder head. Note the special gasket between the distributor housing and the cylinder head, and the self-sealing washer under the bolt head which is located immediately below and to the left of the oil pressure switch (photo).
8 Release all pressure from the rocker arms by using one of the methods described in Section 12.
9 Oil the camshaft bearings and install the camshaft but do not push it fully home until the thrust plate is fitted.
10 Fit the camshaft thrust plate and securing bolts. Should the camshaft endfloat exceed 0.0005 in (0.127 mm) renew the thrust plate.

45 Cylinder head - installation

1 Ensure that the cylinder head and block surfaces are absolutely clean. Clean out the bolt holes, any oil left in them could create enough hydraulic pressure to crack the block when the bolts are screwed in.
2 Obtain a new gasket for your particular engine type (they are not interchangeable) and compare it with the original before discarding the old one.
3 Coat both sides of the new gasket (around the area of the timing cover mating surface only) with jointing compound and position the gasket correctly on the cylinder block.
4 Turn the crankshaft so that all the pistons are positioned a little way down the bores (this will prevent any valve heads damaging the piston crowns until the timing is correctly set later). Lower the cylinder head into place and insert the securing bolts and then tighten them progressively to the specified torque setting in the sequence shown. Fit the oil distribution pipe and tighten its single hollow bolt (photos).
5 Install the camshaft sprocket, as described in Section 42, paragraph 9. It is essential that the mark on the camshaft sprocket flange is correctly aligned otherwise the valves can damage the crowns of the pistons.
6 Before installing the rocker cover, adjust the valve clearances as described in Section 46.
7 Install the distributor, as described in Chapter 4, Section 5.

Fig. 1.23. Cylinder head bolt tightening diagram

42.1 Crankshaft sprocket and key

42.3a Timing chain guides

42.3b Timing chain guide securing circlip

42.6 Installing oil pump chain and sprocket

42.7a Installing timing lower cover

42.7b Essential engagement of chain guide with stop with cover

42.8a Installing crankshaft pulley. Note timing marks on inner and outer rims

42.8b Tightening crankshaft pulley nut

42.9a Installing camshaft sprocket

42.9b Reverse face of camshaft sprocket showing dowel pin

42.12a Installing timing chain tensioner plunger

42.12b Installing timing chain tension spring and plug

43.1 Installing new sump gasket

44.3a Fitting a valve spring lower cup

44.3b Fitting a valve stem oil seal

44.3c Fitting a valve spring

44.5 Installing a rocker shaft and components

44.6a Alignment of rocker shaft cut-out with cylinder head bolt hole

44.6b Installing a rocker shaft circlip

44.7 Distributor housing showing oil pressure switch

45.4a Installing cylinder head

45.4b Tightening a cylinder head bolt

45.4c Installing distribution pipe to cylinder head

46 Valve clearances - adjustment

1   The simplest way to adjust the valve clearances is to set no. 1 piston at TDC on its compression stroke, having first removed the spark plugs. To do this, apply a spanner to the crankshaft pulley nut and turn the crankshaft until placing a finger over no. 1 plug hole, compression can be felt being generated. Continue turning the crankshaft until the second mark (TDC) on the inner rim of the crankshaft pulley is in alignment with the timing cover pointer.

2   Adjust the two valve clearances nearest the front of the engine. The left-hand valves are inlet and the right-hand valves exhaust, when viewed from the driver's seat. Both valves have the same clearance (cold) which is between 0.006 and 0.008 in (0.15 to 0.20 mm).

3   To adjust the gap, release the locknut on the rocker arm and insert a thin rod into one of the holes in the eccentric cam. Turn the cam until the appropriate feeler gauge is a stiff sliding fit between the cam and the end face of the valve stem. **On no account check the clearance between the rocker arm slide pads and the lobes of the camshaft** (photo).

4   Retighten the locknut without moving the eccentric cam.

5   Turn the crankshaft pulley nut until no. 3 piston is at TDC on its compression stroke. This can be judged by watching the high points of the camshaft no. 3 lobes. When they are pointing downwards, check the clearances of no. 3 valves.

6   Turn the crankshaft again until the high points of no. 4 camshaft lobes are pointing downwards and check and adjust the clearances of no. 4 valves (those nearest the back of the cylinder head).

7   Turn the crankshaft again until the high points of no. 2 camshaft lobes are pointing downwards and check and adjust the clearances of no. 2 valves.

8   The sequence of adjustment just described follows the firing order of the engine and avoids unnecessary rotation of the crankshaft.

9   Finally check all rocker adjuster locknuts, remove the spanner from the crankshaft and install the rocker cover using a new gasket (photo).

10 Recheck the valve clearances after 600 miles (1000 km).

47 Engine ancillary components - refitting

1   The ancillary components may be fitted to the engine before or after it is installed. In either case, reverse the removal operations of the components listed in Section 10. Adjust the drivebelts, as described in Chapters 2 and 3.

48 Engine/manual gearbox - reconnecting

1   This is a reversal of the separation procedure described in Section 7, except that the clutch driven plate must be centralised, as described in Chapter 5.

49 Engine/automatic transmission - reconnecting before installation

1   Reverse the operations for removal given in Section 8 but observe the following:

a) *Before joining the torque converter housing to the engine, check that the torque converter is fully to the rear. This is apparent if by placing a straight-edge across the mouth of the bellhousing, the tip of the input shaft is below the rim of the bellhousing. If it is not, press the torque converter into full engagement with the oil pump by turning the torque converter at the same time to engage the driving lugs.*

b) *Tighten the driveplate to torque converter bolts to the specified torque.*

50 Engine (carburettor type) with manual gearbox - installation

1   Reverse the operations given for removal in Section 4, but observe the following:

a) *Make sure that the choke outer cable does not project more than 0.59 in (15.0 mm) beyond its securing clip at the carburettor otherwise the choke flap will not close fully and difficult cold starts may occur.*

b) *On cars with an intermediate shaft type clutch operating mechanism, make sure that the bearing support is positioned at right-angles to the centre-line of the engine and that the bearings are packed with grease and the breather holes in the dust excluding boot face downward.*

c) *Use new self-locking nuts on the propeller shaft front coupling.*

d) *Before tightening the centre bearing bolts, hold the bearing 0.08 in (2.0 mm) in the forward direction to apply the necessary preload.*

e) *Before tightening the right-hand engine mounting bolt set the stop to provide a gap of 0.118 in (3.0 mm) as shown.*

f) *Set the clutch free-movement, as described according to model in Chapter 5.*

51 Engine with automatic transmission - installation

1   This is a reversal of removal operations described in Section 5, but also carry out the operations in paragraphs c, d and e of the preceding Section.

2   Check the operation of the selector lever after reconnection.

52 Engine (fuel injection type) - installation

1   Reverse the removal operations, described in Section 6.

46.3 Checking and adjusting a valve clearance

46.9 Installing rocker cover

Fig. 1.24. Torque converter (automatic transmission) fully installed

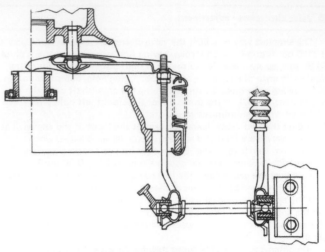

Fig. 1.25. Clutch operating mechanism (intermediate shaft type) setting diagram

Fig. 1.26. Propeller shaft centre bearing preload

A = 0.8 in (2.0 mm) forward pressure

Fig. 1.27. Right-hand engine mounting setting diagram

A = 0.118 in (3.0 mm)

53 Start-up after major overhaul

1   Make sure that the engine, cooling system and transmission have all been refilled with the correct quantity and type of fluid.
2   Check that all controls and leads have been reconnected.
3   Make sure that no spanners or rags have been left within the engine compartment.
4   Increase the engine idling speed slightly to offset the stiffness of the new components.
5   Start the engine in the normal manner. This may take a little longer than normal as fuel has to be drawn from the fuel tank to fill dry components.
6   Once the engine has fired, drive the car for a short distance until normal operating temperature has been reached and then check for oil and water leaks and rectify if necessary.

7   Due to the altered characteristics of the engine by the removal of carbon and valve grinding, the adjustment of the carburettor, ignition, fuel injection and emission control systems should all be checked as described in the relevant Chapters.
8   After the engine has had its first run, allow it to cool completely and check the torque wrench settings of the cylinder head bolts. Similarly, after the first 600 miles (1000 km) check the bolts again **cold**. This time slacken each bolt one quarter of a turn before tightening it to the specified torque. Always tighten the bolts in the sequence given in the diagram (Fig. 1.23).
9   If a number of new engine internal components have been installed, it is recommended that the engine oil and filter are also changed at the end of the first 600 miles (1000 km) running. The engine and road speeds should be restricted for the initial period to assist in bedding-in the new components.

54 Fault diagnosis - engine

Symptom	Reason/s
Engine will not turn over when starter switch is operated	Flat battery Bad battery connections Bad connections at solenoid switch and/or starter motor Defective starter motor
Engine turns over normally but fails to start	No spark at plugs No fuel reaching engine Too much fuel reaching the engine (flooding)
Engine starts but runs unevenly and misfires	Ignition and/or fuel system faults Incorrect valve clearances Burnt out valves Worn out piston rings
Lack of power	Ignition and/or fuel system faults Incorrect valve clearances Burnt out valves Worn out piston rings
Excessive oil consumption	Oil leaks from crankshaft, rear oil seal, timing cover gasket and oil seal, rocker cover gasket, oil filter gasket, sump gasket, sump plug washer Worn piston rings or cylinder bores resulting in oil being burnt by engine Worn valve guides and/or defective valve stem seals
Excessive mechanical noise from engine	Wrong valve to rocker clearances Worn crankshaft bearings Worn cylinders (piston slap) Slack or worn timing chain and sprockets

Note: When investigating starting and uneven running faults do not be tempted into snap diagnosis. Start from the beginning of the check procedure and follow it through. It will take less time in the long run. Poor performance from an engine in terms of power and economy is not normally diagnosed quickly. In any event the ignition and fuel systems must be checked first before assuming any further investigation needs to be made.

Chapter 2 Cooling system

Contents

Specifications

System type	Pressurised, radiator, pump
Capacity	12½ Imp. pints, 7.0 litres, 14.6 US pints
Thermostat opening temperature	176°F (80°C)
Radiator cap pressure rating	13 to 16 psi (0.85 to 1.15 kg/cm^2)

Torque wrench setting	lb/ft	Nm
Water pump securing bolts	20	28

1 General description

1 The cooling system is similar on all models but varies in the design of individual components. The system is pressurised and comprises a radiator, a water pump and fan, a thermostat and the connecting hoses.
2 The layout of the hoses is intricate and incorporates top and bottom radiator hoses, bypass hoses, flow and return hoses for the heater, water jacketed inlet manifold and automatic choke (where fitted).

3 All models except early 1500 versions with a dynamo, have a twin branch water housing bolted to the front left-hand side of the cylinder head into which is screwed the water temperature sender unit.
4 All models except early 1500 models with a dynamo have a sealed type thermostat located at the junction of the radiator top hose, water pump bypass hose and twin branch connector hose (photo).
5 Early 1500 models have a thermostat housing located on the left-hand side of the cylinder head just forward of the fuel pump, instead of the twin branch housing of later models (photo).

1.4a Thermostat location (except early 1500 models)

1.4b Twin branch water housing (except early 1500 models)

Fig. 2.1. Thermostat (T) top hose, (K) bottom hose (W) on early 1500 models

2 Cooling system - draining

1   Should the system have to be left empty for any reason both the cylinder block and radiator must be drained, otherwise with a partly drained system corrosion of the water pump impeller seal face may occur with subsequent early failure of the pump seal and bearing.
2   Place the car on a level surface and have ready a container having a capacity of two gallons which will slide beneath the radiator and sump.
3   Move the heater control on the facia to 'HOT' and unscrew and remove the radiator cap. If hot, unscrew the cap very slowly, first covering it with a cloth to remove the danger of scalding when the pressure in the system is released.
4   Unscrew the radiator drain tap at the base of the radiator and then when coolant ceases to flow into the receptacle, repeat the operation by unscrewing the cylinder block plug located on the right-hand side of the engine. Retain the coolant for further use, if it contains antifreeze.

3 Cooling system - flushing

1   The radiator and waterways in the engine after some time may become restricted or even blocked with scale or sediment which reduce the efficiency of the cooling system. When this condition occurs or the coolant appears rusty or dark in colour the system should be flushed. In severe cases reverse flushing may be required as described later.
2   Place the heater controls to the 'HOT' position and unscrew fully the radiator and cylinder block drain taps.
3   Remove the radiator filler cap and place a hose in the filler neck. Allow water to run through the system until it emerges from both drain taps quite clear in colour. Do not flush a hot engine with cold water.
4   In severe cases of contamination of the coolant or in the system, reverse flush by first removing the radiator cap and disconnecting the lower radiator hose at the radiator outlet pipe.
5   Remove the top hose at the radiator connection end and remove the radiator, as described in Section 7.
6   Invert the radiator and place a hose in the bottom outlet pipe. Continue flushing until clear water comes from the radiator top tank.
7   To flush the engine water jackets, disconnect the hoses from the twin branch connector on the left-hand side of the cylinder head. Place a cold water hose in one branch and let the water flow until it emerges clear from the other branch. On early 1500 models disconnect the top and bottom radiator hoses.

4 Cooling system - filling

1   Place the heater control to the 'HOT' position.
2   Screw in the radiator drain tap. Close the cylinder block drain tap.

3   Pour coolant slowly into the radiator so that air can be expelled through the thermostat pin hole without being trapped in a waterway.
4   Fill to the correct level which is 1 in (25.4 mm) below the radiator filler neck and replace the filler cap.
5   Run the engine, check for leaks and recheck the coolant level.

5 Antifreeze mixture

1   The cooling system should be filled with fresh antifreeze solution every two years. The heater matrix and radiator bottom tank are particularly prone to freeze if antifreeze is not used in air temperatures below freezing. Modern antifreeze solutions of good quality will also prevent corrosion and rusting and they may be left in the system to advantage all year round, draining and refilling with fresh solution each year.
2   Before adding antifreeze to the system, check all hose connections and check the tightness of the cylinder head bolts as such solutions are searching. The cooling system should be drained and refilled with clean water as previously explained, before adding antifreeze.
3   The quantity of antifreeze which should be used for various levels of protection is given in the table below, expressed as a percentage of the system capacity.

Antifreeze volume	Protection to	Safe pump circulation
25%	−26°C (−15°F)	−12°C (10°F)
30%	−33°C (−28°F)	−16°C (3°F)
35%	−39°C (−38°F)	−20°C (−4°F)

4   Where the cooling system contains an antifreeze solution any topping-up should be done with a solution made up in similar proportions to the original in order to avoid dilution.

6 Radiator - removal, inspection, cleaning and refitting

1   Drain the cooling system as described in Section 2.
2   Disconnect the top hose from the radiator header tank and remove the pre-heater housing.
3   Disconnect the bottom hose from the radiator outlet pipe.
4   Unscrew and remove the retaining bolts which secure the radiator to the front engine compartment mounting panel.
5   Lift out the radiator, taking care not to damage the cooling fins. Do not allow antifreeze solution to drop onto the bodywork during removal as damage may result.
6   Radiator repair is best left to a specialist but minor leaks may be tackled with a proprietary compound.
7   The radiator matrix may be cleared of flies by brushing with a soft brush or by hosing.
8   Flush the radiator, as described in Section 3 according to its degree of contamination. Examine and renew any hoses or clips which have deteriorated.
9   Examine the drain tap and its washer, renewing if suspect.
10 Replacement of the radiator is a reversal of the removal procedure. Refill and check for leaks, as described in Section 4.

7 Thermostat - removal, testing and refitting

1   A faulty thermostat can cause overheating or slow engine warm up. It will also affect the performance of the heater.
2   Drain off enough coolant through the radiator drain tap so that the coolant level is below the thermostat housing. A good indication that the correct level has been reached is when the cooling tubes are exposed when viewed through the radiator filler cap.
3   Disconnect the three hoses from the thermostat and remove it (later models) or unbolt the thermostat housing cover (early 1500 models).
4   On early 1500 models withdraw the thermostat cover sufficiently to permit the thermostat to be removed from its seat in the cylinder head.
5   To test whether the unit is serviceable, suspend the thermostat by a piece of string in a pan of water being heated. Using a thermometer, with reference to the opening and closing temperature in Specifications, its operation may be checked. The thermostat should be renewed if it

is stuck open or closed or it fails to operate at the specified temperature (176°F - 80°C). To measure the opening of the thermostat on later models, insert a steel rule, as shown, when cold and again after heating, the difference in reading taking the end of the thermostat hose connecting nozzle as the datum line should be between 0.315 and 0.354 in (8.0 and 9.0 mm). Movement of the thermostat is not instantaneous, give time for it to operate. Never refit a faulty unit.

8 Water pump - description

1 The water pump is of light alloy construction, incorporating an impeller and it is belt-driven from the crankshaft pulley.
2 The water pump can be repaired, if leaking, as described in the following Sections, but where the necessary extractor and pressing facilities are not available it will be advisable to renew it.

9 Water pump - removal and installation

1 Drain the cooling system and remove the radiator, as described earlier in this Chapter.
2 Flatten the tabs of the locking plates and unscrew and remove the fan securing bolts and withdraw the fan.
3 Slacken the dynamo or alternator mountings and adjuster, push the dynamo or alternator in towards the engine and slip off the driving belt and water pump pulley.
4 Disconnect the hoses from the water pump and unscrew and remove the securing bolts.
5 Withdraw the water pump. If it is stuck, do not lever it off or the mating flanges will be damaged but give it a sharp blow with a soft-faced mallet (photo).
6 Installation is a reversal of removal but always use a new flange gasket and copper sealing washers under the bolt heads.

10 Water pump - overhaul

1 Using a suitable two-legged puller, draw the pulley mounting flange from the water pump shaft.
2 Extract the circlip and spacer ring which are now exposed.
3 Support the water pump housing and apply pressure to the end of the shaft in the centre of the impeller. This will push the shaft from the impeller and the shaft/bearing assembly from the water pump housing.
4 Drive out the water pump seal and cover ring from the water pump housing. Fit a new seal.
5 To install the other components, press the shaft/bearing into the housing and then press the impeller onto the shaft using Loctite to secure it. There must be a clearance between the face of the impeller and the housing of between 0.031 and 0.047 in (0.8 and 1.2 mm).
6 Press the flange onto the shaft so that the distance between the front face of the flange and the cylinder block mating face of the pump housing is between 2.957 and 2.973 in (75.1 and 75.5 mm).

11 Fan blades - variation

Dependant upon operating and climatic conditions, the standard four-blade version. This type of fan can be supplied having blades of extra length.

12 Fan belt - adjustment and renewal

1 One or more drivebelts may be employed, according to car model and equipment. A single belt is used to drive the alternator and water pump/fan from a single crankshaft pulley on carburettor models not equipped with emission control systems.
2 Adjustment or renewal of this kind of belt is carried out by loosening the dynamo alternator mountings and adjuster and pushing in towards the engine or pulling away the dynamo alternator until the centre of the top run of the belt can be depressed with the thumb 0.4 in (10.0 mm). Retighten all bolts after adjustment.
3 *Where an air pump is used* (emission control system) a double crankshaft pulley is fitted together with a secondary 'V' belt. Adjustment of the air pump drivebelt, is described in Chapter 3.
4 *On fuel injection models,* two toothed type belts are used, one to drive the alternator and water pump/fan and the other to drive the injection pump. No adjustment is required to these belts but to renew them, refer to Chapter 3.

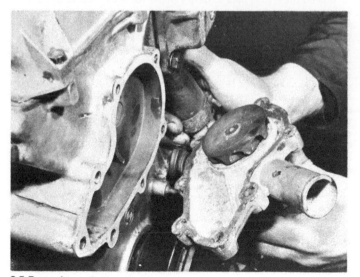

9.5 Removing waterpump

Fig. 2.2. Pre-heater housing at rear of radiator

Fig. 2.3. Testing opening of thermostat with steel rule

Fig. 2.4. Withdrawing water pump pulley mounting flange

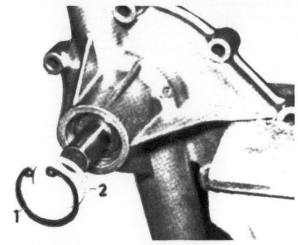

Fig. 2.5. Waterpump shaft/bearing retaining circlip (1) and ring (2)

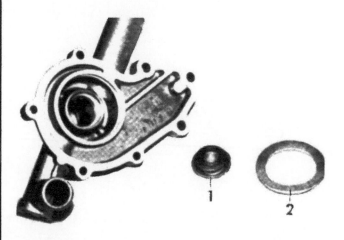

Fig. 2.6. Waterpump seal (1) and cover ring (2)

Fig. 2.7. Water pump impeller to housing clearance (B)

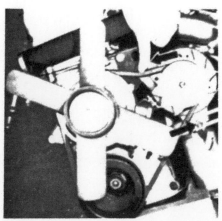

Fig. 2.8. Checking deflection of waterpump - alternator drive belt

13 Fault diagnosis - Cooling system

Symptom	Reason/s
Overheating	Low coolant level
	Slack fan belt
	Thermostat not operating
	Radiator pressure cap faulty or of wrong type
	Defective water pump
	Cylinder head gasket blowing
	Radiator core clogged with flies or dirt
	Radiator blocked
	Binding brakes
	Bottom hose or tank frozen
Engine running too cool	Defective thermostat
	Faulty water temperature gauge
Loss of coolant	Leaking radiator or hoses
	Cylinder head gasket leaking
	Leaking cylinder block core plugs
	Faulty radiator filler cap or wrong type fitted

Chapter 3 Fuel and exhaust systems

Contents

Specifications

Fuel pump (mechanical)

Type ... ...   Pieburg driven by rod from camshaft eccentric

Pressure at 4000 rpm:

1500/1600 ... ...   4.12 lb/in^2 (0.29 kp/cm^2)

2000 series ... ...   4.27 lb/in^2 (0.30 kp/cm^2)

Fuel tank capacity ... ...   10.1 Imp. gals (46.0 litres, 12.2 US gals)

Carburettor - application

1500	Solex 36 PDSI
1600	Solex 36 PDSI
1502	Solex 38 PDSI
1602	Solex 38 PDSI
2000 Touring	Solex 40 PDSI (automatic transmission 40 PDSIT)
2002	Solex 40 PDSI (automatic transmission 40 PDSIT)
2002 TI	Twin Solex 40 PHH
2002 with emission control	Solex 32/32 DIDTA

	36 PDSI	38 PDSI	40 PDSI
Main jet	135 (1500 cc) 140 (1600 cc)	130 140 (1502)	155
Air correction jet	110 (1500 cc) 100 (1600 cc)	100 (1602) 120 (1502)	130
Venturi diameter	26 mm	26 mm	30 mm
Pilot jet	47.5	47.5 (1602) 42.5 (1502)	45
Idling air jet	100	100	100
Accelerator pump tube	80	80	100
Qty. per pump stroke	1.5 to 1.8 cc	1.35 to 1.7 cc	1.8 to 2.2 cc

Fuel inlet valve and gasket thickness	2.0 mm (0.04 in 1.0 mm)	2.0 mm (0.04 in 1.0 mm)	2.0 mm (0.04 in 1.0 mm)
Rich mixture valve	1.0 (1500 cc) 72.5 (1600 cc)	90	100
Fuel level (depth from chamber top flange) ...	0.67 to 0.75 in (17.0 to 19.0 mm)	0.67 to 0.75 in (17.0 to 19.0 mm)	0.67 to 0.75 in (17.0 to 19.0 mm)
Thermostat resistance	—	—	—
Thermostat vent	—	—	—
Thermo start spray nozzle	—	—	—
CO level	—	—	—
	40 PDSIT	**40 PHH**	**32/32 DIDTA**
Main jet	155	130	1st. 117.5 2nd. 137.5
Air correction jet	130	155	1st. 120 2nd. 105
Venturi diameter	30 mm	34 mm	1st. 24 2nd. 28
Pilot jet	45	52.5	—
Idling air jet	100	1.15 to 1.25 mm	—
Accelerator pump tube	100	0.5	50
Qty. per pump stroke	1.75 to 2.05 cc	0.85 to 1.15 cc	0.8 to 1.0 cc
Fuel inlet valve and gasket thickness	2.0 mm (0.04 in 1.0 mm)	2.0 mm (0.04 in 1.0 mm)	2.0 mm (0.04 in 2.0 mm)
Rich mixture valve	100	—	—
Fuel level (depth from chamber top flange) ...	0.67 to 0.75 in (17.0 to 19.0 mm)	—	—
Thermostat resistance	20 ohms	—	—
Thermostat vent	92.5	—	—
Thermo start spray nozzle	50	—	—
CO level	—	—	0.8 to 1.2%

Fuel injection system (2002 TII)

Type	Kugelfischer PL 04-1240181
Pump initial filling	100 cc engine oil
Fuel injection pressure	427 to 540 lb/in^2 (30 to 38 kp/cm^2)
Fuel pump (electric adjacent to rear tank)	Bosch
Fuel pump pressure (at 4000 rpm)	28.45 lb/in^2 (2.0 kp/cm^2)

Slow running speeds

1500	650 to 850 rpm
1600	700 to 800 rpm
1502	700 to 900 rpm
1602	700 to 900 rpm
2000 Touring	700 to 900 rpm
2002 (Manual gearbox)	700 to 900 rpm
2002 (Automatic transmission)	700 to 900 rpm
2002 (Emission control system)	850 to 950 rpm
2002 TI	700 to 900 rpm
2002 TII	850 to 950 rpm

Torque wrench settings

	lb/ft	Nm
Carburettor to inlet manifold	12	16
Fuel pump to rocker cover	9	12
Fuel injection pump cap nut	19	26
Fuel injection pump pulley nut	28	39
Fuel injection valve	23	32

1 General description

1 The fuel system comprises a rear mounted fuel tank, a mechanically operated fuel pump (carburettor models) or an electric type fuel pump (fuel injection versions). Carburettors may be single or twin and there are a number of different types.

2 On cars operating in North America, an exhaust emission control system is installed and a fuel evaporative emission control system.

2 Air cleaner - servicing

1 *On single carburettor models,* pull the hose from the intake pre-heater, disconnect the vacuum hose and slacken the air cleaner body strut nuts.

2 Withdraw the air cleaner from the carburettor and disconnect the crankcase breather hose.

3 Release the cover clips and extract the element. The maximum operating mileage for an element is 10000 miles (16000 km) at which time it should be renewed. Renew more frequently in dusty conditions (photo).

4 Clean out the interior of the air cleaner casing before installing the new element and refitting the assembly which is a reversal of removal and dismantling.

5 *On twin carburettor models,* remove the intake scoop at the rear of the carburettor, pull off the breather hose and loosen the connecting hose clamps at the carburettor.

6 Release the air cleaner securing bolts and remove the assembly.

7 Release the end cover clips, extract the element and renew it, as described in paragraphs 3 and 4.

8   *On fuel injection models,* release the hose clamp, unscrew the two upper mounting bolts and slacken the two lower ones, lift away the air cleaner assembly and detach the crankcase breather hose.
9   Release the end cover clips, extract the element and renew it, as described in paragraphs 3 and 4.

2.3 Air cleaner element (single carburettor)

Fig. 3.1. Air cleaner (twin carburettor)

Fig. 3.2. Air cleaner (fuel injection)

3   Air intake pre-heat valve

1   The valve is located to the right rear of the radiator. With the valve lever in the horizontal position, air drawn in from the front of the car is mixed with air heated by the exhaust manifold in specified proportions by the action of a bi-metal coil which is dependent upon the prevailing engine and ambient temperatures.
2   With the valve lever in the vertical position (summer setting) air is drawn only from outside the car.
3   Periodically remove the cover screw and withdraw the cover. Oil the valve pivot points and check the adjustment.
4   The distance between the valve plate and the rim of the cover (dimension 'A' Fig. 3.3) should be maintained at 2.362 in (60.0 mm). To adjust, release the nuts in the centre of the bi-metal coil.

Fig. 3.3. Setting diagram for pre-heat valve

1   *Locknuts*
A = *2.362 in (60.0 mm)*

4   Fuel filter - renewal

1   An in-line fuel filter is located at the left-hand side of the radiator. This sealed unit should be renewed every 40,000 miles (60000 km).
2   Make sure that the filter is fitted the correct way round - inlet at the bottom and outlet at the top. The filter is usually marked 'IN' (EIN) 'OUT' (AUS).

5   Fuel pumps - description

1   Engines with carburettors employ a mechanically operated fuel pump. This is mounted on the left-hand side of the rocker box and is actuated by a short pushrod from the camshaft.
2   The design of the pump is slightly different according to engine capacity and the differences are shown in the operations described in the next Sections.
3   Engines with a fuel injection system have an electrically-operated pump in conjunction with a small auxiliary tank mounted underneath the car, adjacent to the differential unit. This type of pump starts to operate as soon as the ignition key is switched on and can be heard 'clicking' before the engine is started.

6   Fuel pumps - routine servicing

1   At the intervals specified in 'Routine Maintenance' clean the filter screen of the mechanical type pump.
2   The screen is accessible after removal of the cover centre bolt, cover and gasket on 1500 and 1600 series models. On the larger engines, unscrew the drain plug to which the filter is attached (photo).
3   On electric pumps, loosen the clip which secures the inlet hose to the small auxiliary tank and clean the filter which is incorporated in the end of the hose.

Fig. 3.4. Removing cover from 1500/1600/1602 fuel pump

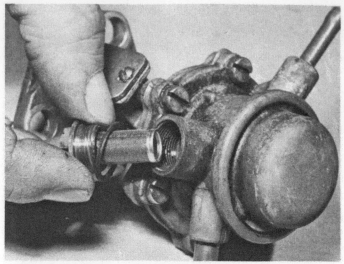

6.2 Drain plug/filter on 2000/2002 fuel pump

7 Fuel pump (mechanical) - testing, removal, overhaul and refitting

1   Disconnect the outlet hose from the fuel pump.
2   Disconnect the distributor LT lead or the central HT lead from the coil to prevent the engine firing and then turn the ignition key to actuate the starter. Observing the outlet nozzle of the pump, well defined spurts of fuel should be ejected. If this is not the case and the cover and inlet hose of the pump are secure, the pump must be overhauled.
3   To remove the pump, disconnect both hoses from the pump and remove the securing bolts (photo).
4   Remove the thick insulator and two thin gaskets and withdraw the operating rod (photo).
5   Scratch a line across the edges of the upper and lower pump flanges so that they can be reassembled in their original relative positions.
6   Remove the cover centre bolt, the cover, gasket and filter (1600 models) or drain plug (2000 series).
7   Remove the flange screws and separate the two halves of the pump.
8   To renew the diaphragm/rod assembly, remove the small cover plate (two screws) extract the operating arm pivot pin circlip and remove the pivot pin. Withdraw the arm and the diaphragm/spring assembly (photo).
9   Renew any worn components. On 2000/2002 type pumps the valves in the pump upper body are renewable but in the pump fitted to engines of smaller capacity, they are not and a fault in their operation must be overcome by renewal of the pump complete.
10 Reassembly is a reversal of dismantling. Tighten the pump flange screws only when the operating arm is held in the depressed (inward) position. Make sure that the upper and lower body scratch marks are in alignment and that the diaphragm is not twisted.
11 Renew the pump cover gasket and washer located under the centre bolt.
12 Refitting is a reversal of removal but renew the thin gaskets which are located one each side of the flange insulator. On no account increase the overall thickness of these parts or the correct operation of the actuating rod and arm will be upset.

Fig. 3.5. Extracting filter from 2000/2002 fuel pump

8 Fuel pump (electric) - removal, overhaul and refitting

1   Disconnect the lead from the battery negative terminal.
2   Disconnect the two pin connector plug from the pump.
3   Disconnect the fuel hoses, plugging the one which runs from the main fuel tank.
4   Disconnect the pump mounting nuts and withdraw the pump complete with auxiliary tank.
5   Disconnect the fuel pump from the auxiliary tank.
6   Any fault with the pump will necessitate renewal of the pump complete as it is a sealed unit.
7   Refitting is a reversal of removal but make sure that the plug is connected the correct way round to the negative and positive terminals on the pump.

Fig. 3.6. Electric fuel pump installed with fuel injection systems

7.3 Removing fuel pump

7.4 Removing fuel pump operating rod

7.8 Operating arm cover plate on fuel pump

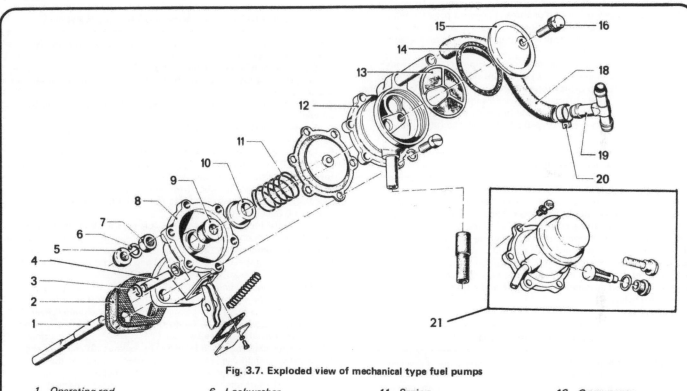

Fig. 3.7. Exploded view of mechanical type fuel pumps

1	Operating rod	6	Lockwasher	11	Spring	16	Cover screw
2	Insulator	7	Nut	12	Upper body	18	Hose
3	Circlip	8	Lower body	13	Screen	19	Tee connector
4	Pivot	9	Collar	14	Gasket	20	Clip
5	Insulating bush	10	Retainer	15	Cover	21	Alternative type pump housing and cover

Fig. 3.8. Electric fuel pump connector

Fig. 3.9. Electric fuel pump mounting

9 Fuel tank - removal and installation

1   Drain the fuel from the tank either by unscrewing the drain plug or if one is not fitted by syphoning it out with a length of tubing.
2   Disconnect the lead from the battery negative terminal.
3   Remove the floor panel from the luggage boot and disconnect the leads from the tank transmitter unit (photo).
4   Disconnect the hose from the pipe on the transmitter unit. On fuel injection models, two hoses are connected to the tank transmitter unit.
5   Release the clip at the base of the filler tube and push the rubber sleeve upwards (photo).
6   Remove the fuel tank securing bolts and lift the tank upwards from its mounting.
7   It should be noted that on 2000 Touring models, the side pocket within the luggage area will have to be removed before the floor panel can be withdrawn.
8   If the tank is to be cleaned of sediment it will usually be adequate to shake it vigorously using two or three changes of paraffin and finally washing it out with clean fuel. Before shaking the tank it will be advisable to remove the transmitter unit using two crossed screwdrivers engaged in the rim cut-outs.
9   If there is a leak in the tank make only temporary repairs with fibreglass or similar material. Any permanent repairs by soldering or welding must be left to professionals due to the risk of explosion if the tank is not first purged of fumes by steam cleaning.
10 Installation is a reversal of removal but before inserting the transmitter unit, clean the filter gauze and renew the rubber sealing ring (photo).

10 Carburettors - general description

1   A number of different carburettors have been used according to engine type, date of production and operating territory. For precise application reference should be made to 'Specifications' Section.
2   In general terms, all carburettors are of Solex manufacture, early models having manually operated chokes while cars equipped with automatic transmission or emission control systems have a coolant heated automatic choke. Later models with emission control systems use a carburettor with an electrically-pre-heated choke in addition to the coolant-heated automatic type choke and a float bowl pressure relief return valve.

11 Solex PDSI series carburettors - slow running adjustment

1   Run the engine until normal operating temperature is reached. Make sure that the choke is fully off.
2   Ensure that the ignition timing and valve clearances are correct.
3   If the carburettor has been dismantled or is grossly out of tune, screw the mixture screw in as far as it will go (do not force it into its seat) and then unscrew it 2½ complete turns to provide an initial setting.
4   Turn the throttle stop screw until the engine is running at between 700 and 800 rpm.
5   Now turn the mixture screw in or out until the fastest and smoothest speed is obtained. If necessary reduce the idling speed to that specified by adjusting the throttle stop screw slightly.
6   This method must be regarded as far from precise and adjustment for maximum economy and performance should be carried out with a tuning device such as a Colortune, a vacuum gauge or a CO meter where these are available.

12 Solex 40PHH (twin) carburettors - slow running adjustment

1   Have the engine at normal operating temperature with the ignition timing and valve clearances correctly set.
2   Refer to Fig. 3.12. Gently tighten the four mixture screws (1, 2, 3, 4) and then unscrew them ½ a turn each.
3   Unscrew the synchronising screw (5) until it is no longer in contact with the throttle lever.
4   Unscrew the throttle stop screw (6) as far as possible and then screw in the synchronising screw (5) until it just makes contact with the throttle lever.

5   Now screw in the throttle stop screw (6) until it just makes contact and then screw it in a further two complete turns.
6   The use of a carburettor balancing device will now be required.
7   Remove the air cleaner, start the engine and adjust the engine speed to 1200 rpm by turning the throttle stop screw (6).
8   Using the balancing device (or airflow meter), synchronise the airflow of carburettor intakes 2 and 3 by adjusting screw (5). The intakes are regarded as 1 to 4 numbering from the front of the engine.
9   Synchronise intake 1 with 2 by adjusting screw (7).
10 Synchronise intake 3 with 4 by adjusting screw (8).
11 When the synchronising is completed, adjust mixture screws (1) to (4) for smoothest running and set the throttle stop screw (6) to give an idling speed of between 700 and 900 rpm.
12 If the period of adjustment is somewhat lengthy, occasionally rev. up the engine to prevent fouling of the spark plugs.
13 Switch off the engine and refit the air cleaner.

13 Solex 32/32 DIDTA carburettor - slow running adjustment

1   Have the engine at normal operating temperature with the ignition timing and valve clearances correctly set.
2   Check the adjustment of the dashpot. To do this, have the engine running at idling speed and pull the vacuum pipe from the dashpot on the carburettor.
3   Plug the open end of the vacuum pipe and release the dashpot locknut.
4   Screw the dashpot in, or out, until the engine idling speed is between 1550 and 1650 rpm.
5   Unplug the vacuum pipe and reconnect it to the dashpot.
6   Now set the engine idling speed to between 850 and 950 rpm by turning the carburettor air bypass screw (2).
7   A CO meter must now be attached to the exhaust pipe and the mixture regulating screw (3) turned until the CO volume ranges between 0.8 and 1.2%.
8   Should the idling speed fluctuate, it is possible that the throttle butterfly valve opening is incorrectly set. Refer to Fig. 3.14. To adjust this, gently tighten screw (2) and then turn screw (4) in or out until the engine idling speed is between 650 and 700 rpm. Now turn screw (3) until the CO reading is approximately 3%.
9   Release screw (2) and turn it until the CO volume is between 0.8 and 1.2% and the engine idling speed is between 850 and 950 rpm.
10 Having set the first stage (primary) throttle valve plate opening, now adjust the second stage. To do this, switch off the ignition and remove the carburettor (see Section 17).
11 Invert the carburettor and release the locknut (6) and turn the screw (5) until light can just be seen between the edge of the valve plate and the carburettor wall. Retighten the locknut.
12 Should the fast idle speed be incorrect, the next time that the engine is started from cold, first run it to normal operating temperature, switch off, remove the air cleaner assembly and then pull the vacuum pipe from the distributor.
13 Slightly raise the accelerator linkage arms so that the choke butterfly can be closed with the fingers but leaving a gap of 0.1181 in (3.0 mm) - use a twist drill for this - and at the same time move the follower to the second step on the fast idle cam.
14 Re-start the engine when the fast idle speed should be between 2300 and 2500 rpm. Any correction should be carried out by releasing the locknuts and moving the position of the arm adjacent to the dashpot connecting rod.

14 Solex PDSI series carburettor (manual choke) - removal and installation

1   Remove the air cleaner.
2   Disconnect the fuel inlet hose.
3   Disconnect the choke cable and accelerator linkage (photo).
4   Disconnect the distributor vacuum hose.
5   Unscrew the two carburettor flange screws and lift the carburettor from the inlet manifold and lift off the flange gaskets.
6   Installation is a reversal of removal but make sure that the choke outer cable does not project more than 0.6 in (15.0 mm) beyond its clamp towards the front of the car otherwise the choke valve plate will not close fully.

9.3 Fuel tank transmitters

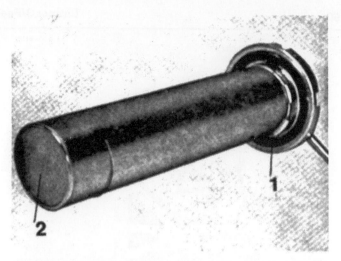

Fig. 3.10. Fuel tank transmitter unit filter (2) and sealing ring (1)

9.5 Fuel tank filler pipe

Fig. 3.11. Solex PDSI adjustment screws

1 Throttle stop screw
2 Mixture regulating screw

14.3 Throttle rod return spring (single carburettor)

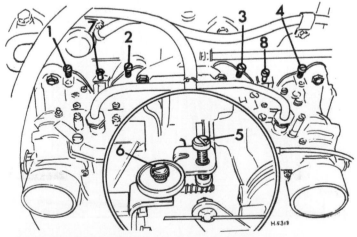

Fig. 3.12. Solex 40 PHH (twin) adjustment screws

1 Mixture regulating screw	5 Synchronising screw
2 Mixture regulating screw	6 Throttle stop screw
3 Mixture regulating screw	7 Connecting shaft screw
4 Mixture regulating screw	8 Connecting shaft screw

Fig. 3.13. Solex 32/32 DIDTA dashpot
1 Locknut

Fig. 3.14. Solex 32/32 DIDTA adjustment screws

2   Air bypass screw   4   Speed screw
3   Mixture regulating screw

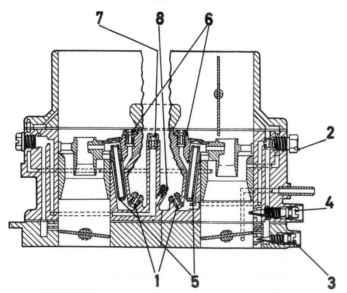

Fig. 3.15. Location of jets and adjuster screws (32/32 DIDTA)

1   Main jet   5   Mixing tube
2   Idling jet   6   Air correction jet
3   Mixture regulating screw   7   Intermediate air jet
4   Air bypass screw   8   Second stage intermediate jet

Fig. 3.16. Solex 32/32 DIDTA secondary throttle butterfly valve

5   Adjustment screw   6   Locknut

15 Solex 40 PDSIT series carburettor (automatic choke) - removal and installation

1   Remove the air cleaner and disconnect the fuel inlet hose.
2   Pull the lead from the thermostat valve.
3   Unscrew and remove the choke housing cover screws.
4   Remove the safety clip from the end of the accelerator shaft and depress the shaft to disengage it from the throttle lever of the carburettor.
5   Pull off the distributor vacuum hose from the carburettor and move the automatic choke housing to one side without disconnecting the coolant hoses from it.
6   Remove the flange securing nuts and lift off the carburettor and its gaskets from the inlet manifold.
7   Installation is a reversal of removal but make sure that the choke bi-metallic spring engages in its notch in the operating arm as the housing cover is offered up. Set the notch on the cover in alignment with the projection on the housing before tightening the securing screws.

16 Solex 40 PHH (twin) carburettors - removal and installation

1   Remove the air cleaner.
2   Disconnect the choke cable.
3   Release the engine oil dipstick retainer clip.
4   Disconnect the fuel inlet hoses from the carburettors.
5   Disconnect the accelerator linkage by detaching the tension spring, loosening the bracket and pressing the pull rod balljoint from the shaft ball.
6   Unscrew and remove the carburettor flange nuts and withdraw the carburettors from the inlet manifold. Remove the flange gaskets.
7   Installation is a reversal of removal but repeat the choke cable setting procedure, as described in Section 14, paragraph 6.
8   Make sure that the torsion spring is installed with its ends as shown in Fig. 3.20.
9   Before attaching the right-hand carburettor to the manifold engage the peg with the hole in the choke butterfly plate lever.

17 Solex 32/32 DIDTA carburettor - removal and installation

1   The operations are similar to those described in Section 15, except that the flange is secured to the manifold with four nuts instead of two.

Fig. 3.17. Solex 40 PDSIT accelerator linkage at carburettor

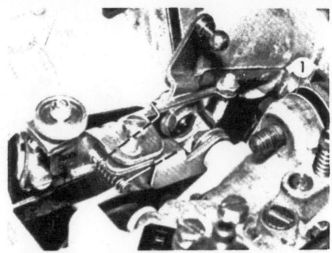

Fig. 3.20. Correct installation of torsion spring at carburettor (Solex 40 PHH)

Fig. 3.18. Installation of automatic choke housing cover on Solex 40 PDSIT

Fig. 3.21. Engaging choke butterfly plate peg (right-hand and 40 PHH carburettor)

Fig. 3.19. Disconnecting accelerator linkage at carburettors (Solex 40 PHH)

18 Solex PDSI series carburettor - overhaul

1   Overhaul of manual and automatic choke versions is very similar, the main difference lying in the automatic choke housing.

2   Having removed the carburettor, clean the external surfaces. With automatic choke versions, the choke housing cover can be removed from the coolant hoses if necessary, provided the radiator cap is removed to release any pressure in the system and the choke connecting hoses are kept at their highest level there is no need to drain the cooling system.

3   On manual choke versions commence dismantling by disconnecting the choke link rod from the throttle starter lever.

4   Remove the screws and lift off the float chamber cover.

5   Unscrew the fuel inlet needle valve and seat.

6   Remove the pivot pin and extract the float.

7   Unscrew the float chamber plug and the main jet.

8   Pull the accelerator pump jet from its location on the top face of the carburettor.

9   Unscrew the idling jet and withdraw the emulsion jet.

10 Remove the accelerator pump cover (4 screws) and the diaphragm and spring.

11 Unscrew and remove the mixture screw and spring.

12 If an altitude corrector is installed, remove the flange valve.

13 This should be the limit of dismantling. Wear in throttle valve plates

1   Gasket
2   Cover
3   Cover screw
4   Choke plate spindle
5   Screw
6   Choke valve plate
7   Roller
8   Circlip
9   Link rod
10 Choke lever assembly
11 Circlip
12 Trunnion
13 Spring
14 Adjuster
15 Locknut
16 Tapered lock screw
17 Nut
18 Accelerator pump assembly
19 Float chamber and
    main body
20 Float
21 Throttle valve housing

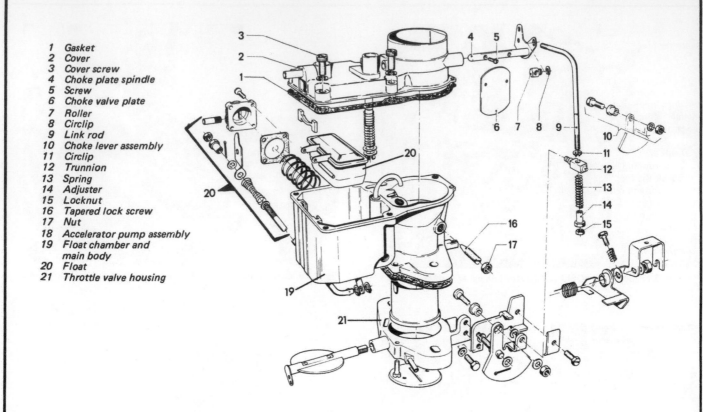

Fig. 3.22. Components of Solex PDSI carburettor

1   Screw
2   Throttle valve housing
3   Gasket
4   Lockwasher
5   Nut
6   Throttle lever
7   Clamping spring
8   Choke link rod
9   Throttle stop screw
10 Spring
11 Locknuts
12 Screw
13 Nut
14 Grub screw
15 Seal
16 Cover
17 Spring
18 Diaphragm
19 Gasket
20 Protecting sleeve
21 Cover bolt
22 Seal
23 Choke housing cover
24 Seal
25 Screw
26 Choke housing
27 Gasket
28 Choke valve plate spindle
29 Securing screws
30 Choke valve plate
31 Return spring
32 Spacers
33 Cover
34 Float
35 Body
36 Accelerator pump assembly

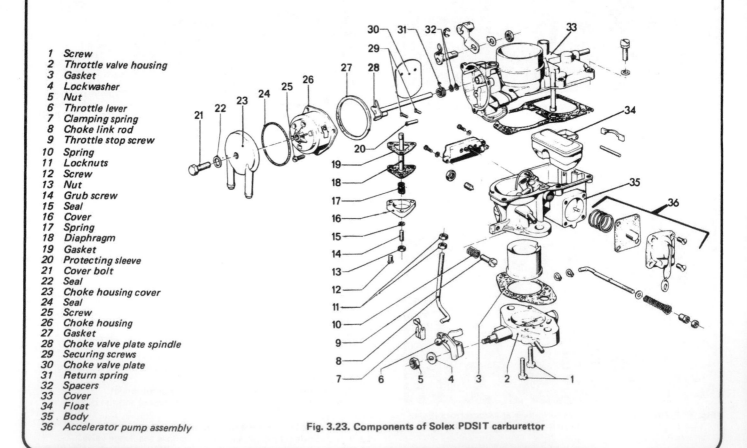

Fig. 3.23. Components of Solex PDSIT carburettor

or spindles will necessitate a new or rebuilt carburettor. Renew any worn or damaged components and obtain a repair kit of gaskets. Clean every part in fuel and blow through jets and passages with air from a tyre pump - never use wire to probe jets. Take the opportunity of comparing calibration markings on jets and other components with those listed in Specifications in case a previous owner has substituted ones of incorrect type.

14 Reassembly is a reversal of dismantling but once reassembled and installed and the engine brought to normal operating temperature, carry out the following adjustments and checks:

Manual and automatic choke carburettors
A Slow-running adjustment (See Section 11).
B Check the fuel level by disconnecting the fuel inlet pipe from the carburettor and removing the cover and gasket. Using a depth gauge check the level of the fuel in the carburettor bowl. This should be ¾ in (19.05 mm) below the top mating flange of the bowl. If adjustment is required, alter the thickness of the fuel inlet valve washer.

Automatic choke (40 PDSIT) carburettor
C To set the fast idle speed (choke fully closed), move the accelerator rod at the carburettor until the choke valve can be closed with the fingers to within 0.25 in (6.5 mm) of the carburettor throat inner wall. The stop lever will now spring into the fast idle position. By means of the locknuts on the choke link rod adjust the fast idle speed to 2100 rpm. To increase fast idle speed, lengthen the rod or to decrease it, shorten the rod.
D Check the automatic choke adjustment by switching off the engine and removing the automatic choke housing cover, but leaving the coolant hoses connected. Insert a screwdriver into the choke housing and depress the choke rod to its stop. At the same time check the clearance between the edge of the choke valve plate and the carburettor throat wall. Use a twist drill for this when the gap should be 0.25 in (6.5 mm). If necessary release the locknut at the base of the choke rod and adjust.
E Refer to Fig. 3.27. Check the thermostart valve by removing the cover of the choke housing. If the temperature is 59°F (15°C) or above, the valve 'A' must be open (lifted from its seat). If the temperature is below 59°F (15°C) switch on the ignition and after one minute the bi-metallic spring 'C' should have raised the valve between 0.04 and 0.08 in (1.0 and 2.0 mm). Where the operation is faulty, check the security of the connecting lead and the bi-metallic spring for a fracture. Do not alter the setting of seat 'B' as this is set in production.

19 Solex 40 PHH (twin) carburettors - overhaul

1 The jets of each carburettor are accessible from outside. Any obstruction should be cleared by applying air pressure to them from a tyre pump.
2 Access to the float is obtained by removing the float chamber cover and gasket.
3 Do not dismantle components unnecessarily and if wear has occurred in the valve plates or spindles, renew the carburettor or obtain a rebuilt one.
4 When the carburettor has been reassembled and installed, carry out the following checks and adjustments.
   a) Adjust the slow-running (Section 12).
   b) Check the float-level. To do this, run the engine for a minute or two and then switch off the ignition and remove the float chamber cover. Using a depth gauge, measure the level of the surface of the fuel below the rim of the cover flange. This should be ¾ in (19.05 mm). The level can be adjusted by releasing the locknut on the fuel inlet valve and adjusting the position of the inlet nozzle having first detached the hose. One complete turn of the nozzle alters the fuel level 0.16 in (4.0 mm).
   c) To check the amount of fuel ejected by the accelerator pump requires the use of a metering device but the direction of spray can be checked and adjusted in accordance with the diagram. Should poor acceleration be experienced with exceptional fuel economy, the fuel pump can be adjusted temporarily (pending the employment of a metering device) by releasing the locknut at the bottom of the pump rod and screwing the adjuster nut upwards. If acceleration is jerky and accompanied by black smoke from the exhaust and heavy fuel consumption, then reduce the pump stroke by unscrewing the nut downwards.

   d) A choke clearance must exist between the end of the stud and the choke lever as shown. Set the pull rod so that the distance between the centres of the eyes is approximately 1.85 in (47.0 mm) to achieve this.
   e) Always set the pull rod located between the carburettor throats to a length (between balljoint centres) of 1.6 in (40.4 mm). Whenever the length of the pull rod is adjusted this must be followed by synchronising the carburettors.

20 Solex 32/32 DIDTA carburettor - overhaul

1 Disconnect the choke link from the throttle lever.
2 Disconnect the dashpot connecting lever.
3 Unscrew and remove the carburettor cover.
4 Unscrew the fuel inlet needle valve from the underside of the cover and extract the sealing washer.
5 The automatic choke housing and thermostart element may be removed from the cover if required.
6 Remove the pivot pin and extract the float from the float chamber.
7 The accelerator pump discharge nozzle is a press fit in the carburettor body and should only be extracted if essential.
8 Remove the primary idle jets, the secondary idle plug and the primary and secondary high-speed air jets.
9 Remove the primary and secondary main metering jets.
10 Extract the split pin from the accelerator pump connecting rod and then remove the pump.
11 Unscrew and remove the mixture screw from the throttle valve body.
12 Remove the idle air bypass screw from the carburettor body.
13 Check all components for wear or damage. Blow through all jets with air from a tyre pump and take the opportunity to compare their calibration marks with those listed in Specifications in case a previous owner has changed them.
14 Obtain a repair kit which will contain all the necessary gaskets and washers.
15 Reassembly is a reversal of dismantling but observe the following:
   a) Make sure that the automatic choke bi-metallic spring engages with the end of the intermediate lever as the housing cover is fitted. With the cover correctly aligned and installed, the choke valve plate should be closed at room temperature.
   b) Adjust the secondary throttle valve plate gap to prevent interlock jamming. To do this, release the locknut on the secondary throttle valve stop screw and unscrew the stop screw until the valve plate is completely closed. Now turn the screw ¼ turn inwards and re-tighten the locknut.
   c) Install the accelerator pump connecting rod in the lower hole of the primary throttle shaft lever while the split pin should be inserted in the outer hole of the pump actuating lever. Make sure that the washers are correctly positioned in relation to the connecting rod spring as shown. With the throttle valve plate completely closed, there should be no clearance between the pump lever and pump plunger rod.
   d) Make sure that the accelerator pump discharge nozzle points into the primary venturi.
   e) No float level adjustment is provided for but the fuel level will be correct if a sealing washer (0.08 in - 1.03 mm) thick, as specified is installed under the fuel inlet needle valve.
   f) Adjust the slow-running as described in Section 13 once the carburettor has been installed to the engine. Initial setting of the mixture control screw should be five turns open.

21 Fuel injection system - description

This system is fitted to 2002 TII models. Fuel is drawn from the tank by an electrically-operated pump, filtered and pumped into the injection pump, which is belt-driven from the crankshaft. The injection pump meters the air and fuel separately into the inlet manifold in the correct proportions according to engine load and speed requirements. This is achieved by the accelerator linkage and a governor, The injection pump camshaft is so designed that fuel is injected in synchronisation with the engine firing order.

An additional feature of the system is a solenoid valve which injects fuel at initial engine starting for a period determined by the coolant temperature.

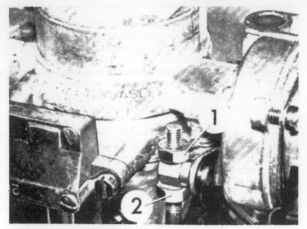

Fig. 3.24. Fast idle adjuster locknuts (1 and 2) on Solex 40 PDSIT

Fig. 3.25. Depressing choke rod (Solex 40 PDSIT)

Fig. 3.26. Choke valve plate adjusting screw and locknut (1) on Solex 40 PDSIT

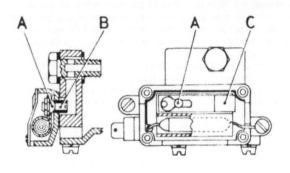

Fig. 3.27. Diagram of thermostat valve (Solex 40 PDSIT)

A   Valve   C   Bi-metallic spring
B   Valve seating

Fig. 3.28. Location of jets in Solex 40 PHH carburettor

1   Main jet carrier
2   Main jet
3   Seal
4   Spring
5   Mixture screw
6   Idling air jet
7   Fuel inlet needle valve
8   Seal
9   Air correction jet
10 Hollow screw
11 Seal
12 Injection tube
13 Seal
14 Seal
15 Ball valve

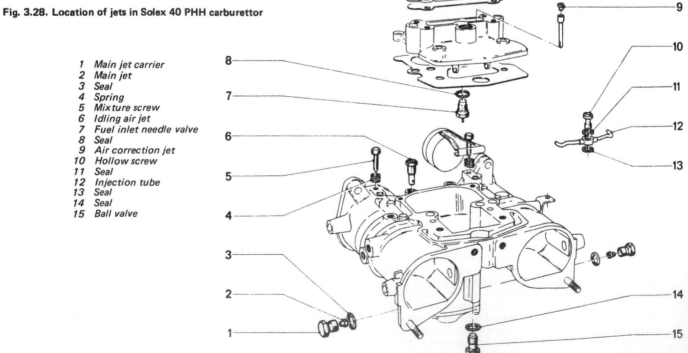

Fig. 3.29. Components of Solex 40 PHH carburettor

1 Screw
2 Cover
3 Diaphragm
4 Screw
5 Pivot
6 Adjusting nut
7 Locknut
8 Arm
9 Spring
10 Washer
11 Diaphragm spring
12 Lower body
13 Diaphragm
14 Nut
15 Adjusting screw
16 Nut
17 Lockwasher
18 Support
19 Spring
20 Spacer
21 Spacer
22 Bush
23 Choke rod lever
24 Throttle lever
25 Throttle valve plate spindle
26 Screw
27 Link rod
28 Circlip
29 Interconnecting lever
30 Torsion spring
31 Interconnecting lever
32 Nut
33 Adjusting screw

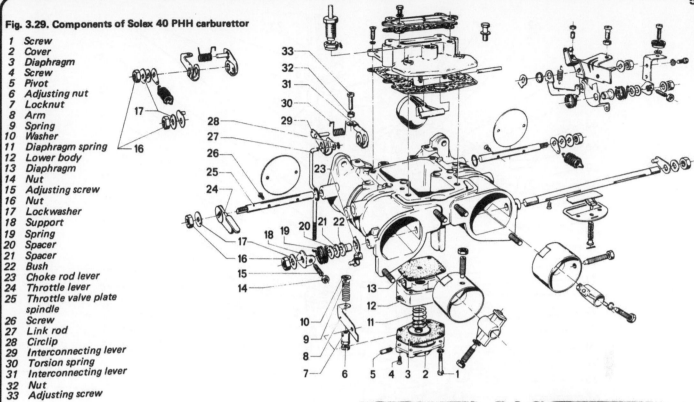

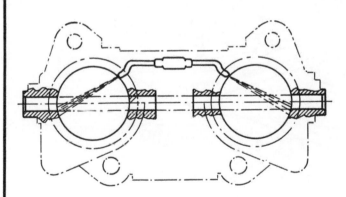

Fig. 3.31. Accelerator pump adjuster (1) and locknut (2) on Solex 40 PHH carburettor

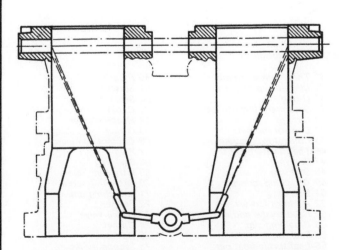

Fig. 3.30. Accelerator pump ejection diagram (Solex 40 PHH)

Fig. 3.32. Choke pull rod setting diagram (Solex 40 PHH)

A = in (0.2 mm)   B = 1.85 in (47.0 mm)

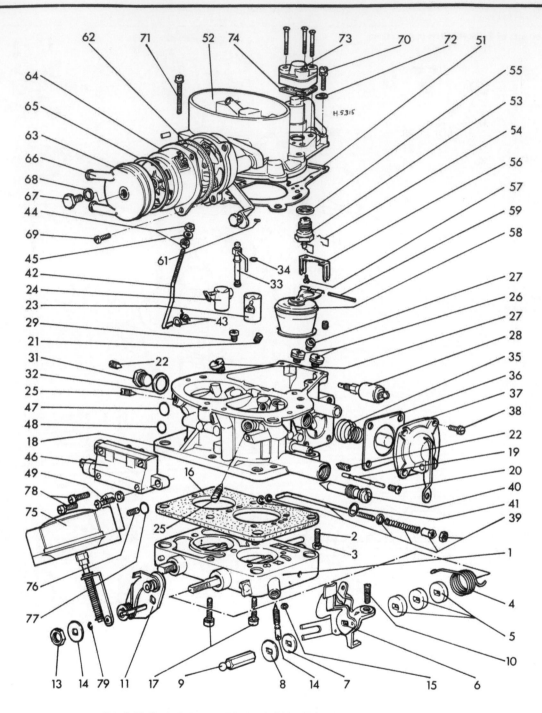

Fig. 3.33 Exploded view of Solex 32/32 DIDTA carburettor

1 Throttle body
2 Throttle valve stud
3 Throttle valve nut
4 Idle air valve spring
5 Shaft spacer bushing
6 Throttle valve lever
7 Valve shaft washer
8 Cam lever washer
9 Valve shaft nut
10 Throttle adjusting screw
11 Transfer lever
12 Valve shaft washer
13 Valve shaft nut
14 Idle mixture screw
15 O-ring
16 Gasket
17 Screws
18 Float bowl and
   carburettor body

19 Carburettor cover pin
20 Carburettor cover screw
21 Idle set
22 Venturi retaining screw
23 Primary venturi
24 Secondary venturi
25 Venturi retaining screw
26 Main jet
27 Air correction jet
28 Idler interrupter vent (bypass)
29 Progression jet
31 Plug screw
32 Seal ring
33 Injection tube
34 O-ring
35 Diaphragm spring
36 Accelerator pump diaphragm
37 Accelerator pump cover
38 Attaching screws

39 Throttle rod assembly
40 Idle air bypass screw
41 O-ring
42 Secondary throttle rod
43 Circlips
44 Control rod nuts
45 Control rod washers
46 Thermo-start device
47 Seal ring
48 Seal ring
49 Attaching screws
51 Gasket
52 Air horn (cover)
53 Needle valve and seat
54 Retaining clip
55 Seal ring
56 Leaf spring float pin
57 Retaining screw
58 Float

59 Float pin
61 Setscrew
62 Seal ring
63 Choke cover
64 Heater element
65 O-ring
66 Coolant connector pipe
67 Plug
68 Seal ring
69 Attaching screw
70 Attaching screw
71 Attaching screw
72 Washer
73 Power valve
74 Gasket
75 Dashpot
76 Secondary jet
77 Seal ring
78 Attaching screw
79 Circlip

Fig. 3.34. Setting diagram for pull rod located between twin 40 PHH carburettors

C = 1.6 in (40.4 mm)

22 Fuel injection system - servicing

1   It is not recommended that the injection pump or other individual components are dismantled but if after reference to the relevant fault diagnosis Section if any part requires renewal, remove and install it, as described in the following Sections.
2   Any adjustment should be limited to the operations described in the next Section.

23 Throttle valve with injection pump (fuel injection) - synchronising

1   As an initial setting, adjust the length (between balljoint centres) of the connecting rod to 3.346 in (85.0 mm).
2   Withdraw the cover from the throttle valve.
3   Release the locknut and unscrew the screw until it is no longer in contact with the cam.
4   Release the rod clamp screws.
5   Using a suitable piece of rod, align the upper slot in the regulating lever with the hole in the injection pump housing and retain it in this position.
6   Insert a rod 0.157 in (4.0 mm) diameter into the hole within the throttle valve housing, press the cam against it and making sure that the connecting rod is free within the clamp, tighten the clamp screws without moving the position of the cam. Remove the rod from the hole in the throttle valve housing and observe that the edge of the cam slightly overlaps the hole.
7   Adjust the slow-running speed to between 850 and 950 rpm using the screw and locknut.

24 Warm-up sensor (fuel injection) - adjustment

1   This adjustment must only be carried out with the engine cold.
2   Remove the air cleaner housing, as described in Section 2, paragraph 9.
3   Using a screwdriver, press out the air regulating cone until a thin strip of metal can be slid into its groove to retain it in the projecting position.
4   Refer to Fig. 3.39. Measure the distance (A) between the end of the screw and stop bolt which should be between 0.090 and 0.114 in (2.3 and 2.9 mm). If this is not the case, adjust the position of plate (1).
5   Refer to Fig. 3.40. Run the engine to normal operating temperature and adjust the idling speed. Under these conditions, the air regulating cone must project 0.39 in (10.0 mm) - dimension (A) and the distance (B) 0.157 in (4.00 mm). The grub screw (1) should be just in contact with the stop bolt (2).

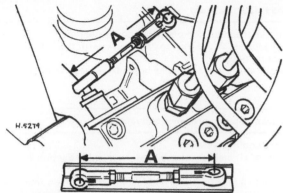

Fig. 3.35. Initial setting of connecting rod (fuel injection system) for synchronising throttle valve and injection pump

A = 3.346 in (85.0 mm)

Fig. 3.36. Throttle valve adjuster screw (1) and connecting rod clamp screws (2) - fuel injection system

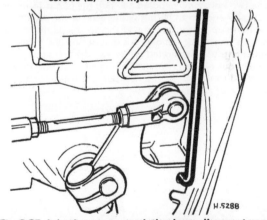

Fig. 3.37. Injection pump regulating lever alignment pending synchronising

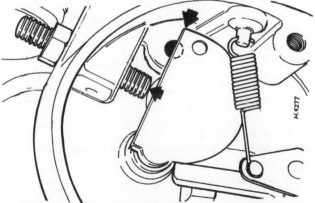

Fig. 3.38. Correct alignment of throttle valve cam with hole - fuel injection system

25 Full load (fuel injection) - setting

1   Disconnect the induction pipe from no. 1 cylinder.
2   Align the lowest slot in the regulating lever with the hole in the injection pump housing.
3   Adjust the stop screw so that its head just makes contact with the lever.

26 Fuel injection pump - removal and refitting

1   Drain the cooling system.
2   Remove the air cleaner assembly, as described in Section 2, paragraph 9.
3   Disconnect all injection lines and fuel line banjo union from the pump.
4   Disconnect the fuel return hose, oil feed hose, coolant inlet hose and the combined dipstick tube mounting with starter cable eye.
5   Disconnect the coolant return hose, the oil return hose and the air warming-up hose. Unscrew the bolt (7) shown in Fig. 3.44.
6   Disconnect the link rod from the injection pump lever and detach the cover (four bolts) from the toothed drivebelt.
7   Turn the crankshaft by applying a spanner to the crankshaft pulley nut until no. 1 piston is at TDC. This can be established by removing no. 1 spark plug and with a finger placed over the hole feeling the compression being generated. As soon as this is felt, watch the notch in the crankshaft pulley and when this is in alignment with the mark on the front of the lower timing cover the TDC position has been reached. At this stage check also that the notch in the toothed driving belt would be in alignment with the projection on the belt cover (if the latter were fitted), also that the mark on the front face of the pump sprocket is in alignment with the mark on the flange of the drivebelt housing.
8   Unscrew and remove the pump sprocket nut and draw off the sprocket with a suitable extractor. Disengage the sprocket from the belt.
9   Remove the two bolts now exposed and pull the pump from the drivebelt housing far enough to permit withdrawal of the intermediate shaft from the warm-up sensor housing. Withdraw the pump completely.
10 Installation is a reversal of removal but do not turn the crankshaft while the components are dismantled, otherwise the belt and sprocket will be completely out of alignment. Make sure all hoses and pipes are correctly connected to the pump. If a new pump is fitted it must be charged with 100 cc of engine oil.

Fig. 3.40. Correctly set warm-up sensor on fuel injection system

1   Grub screw   A = 0.39 in (10.0 mm)
2   Stop bolt   B = 0.157 in (4.0 mm)

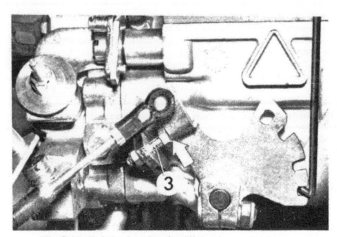

Fig. 3.41. Full load setting (fuel injection system)

3   Stop screw

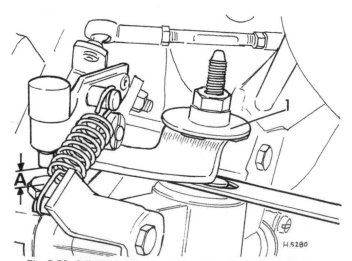

Fig. 3.39. Adjusting warm-up sensor on fuel injection system

1   Plate
A = 0.090 to 0.114 in (2.3 to 2.9 mm)

Fig. 3.42. Fuel injection pump connections with cylinders numbered 1 to 4 from front of engine. Note filter (S)

Fig. 3.43. Fuel return hose (1) oil feed hose (2) coolant inlet hose (3) A is position of clip screw essential to prevent interference with enrichment lever

Fig. 3.44. Coolant return hose (4) oil return pipe (5) warm-up air hose (6) securing bolt (7) on fuel injection sensor housing

Fig. 3.45. Fuel injection pump drivebelt cover bolts

8 Connecting link

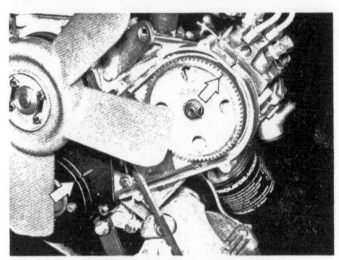

Fig. 3.46. Fuel injection pump sprocket alignment mark

Fig. 3.47. Disconnecting intermediate shaft from fuel injection warm-up sensor housing

27 Fuel injection pump drivebelt - renewal

1 Carry out the operations described in paragraphs 2 and 7, of the preceding Section.
2 Loosen the alternator mounting and adjustment strap bolts and slip the notched alternator driving belt from the pulleys.
3 Remove the crankshaft pulley (four bolts).
4 Slacken the topmost toothed belt cover bolt and remove the lower ones. Slip the belt from the pump sprocket and then pull the belt cover forward so that the belt can be withdrawn downwards and out of the housing.
5 Refitting is a reversal of removal but on no account turn the crankshaft while the injection pump belt is removed otherwise the belt and sprocket cannot be aligned correctly.
6 The notched type alternator belt requires no adjustment.

28 Start valve/thermo-time switch (fuel injection) - testing

1 *To test the start valve,* first detach it from the throttle valve section and then switch on the ignition.
2 Connect a piece of electrical cable between the battery positive terminal and the 'SV' terminal of the time switch. If fuel is immediately ejected, the valve is in good order. As soon as the temporary lead is

removed no dripping should occur from the valve. Refit the valve using a new 'O' ring if necessary.

3   *To test the thermo-time switch,* pull the plug from it and then connect a test lamp between the battery positive terminal and terminal 'W' on the thermo-time switch. The switch is in good order if the test lamp lights up at a coolant temperature below approximately 92°F (31°C). With the test lamp still connected, use another lead between the battery positive terminal and terminal 'G' on the switch. After a short interval, the bi-metallic contact will open and the test lamp should go out.

4   To check the time-switch, unscrew the switch from the engine compartment rear bulkhead. Connect a test lamp between the 'SV' terminal and earth. Pull the distributor LT lead from the coil and then actuate the starter motor. After a short interval, the test lamp should go out, the interval is dependent upon the coolant temperature:

    4°F   (−20°C) − 9 to 15 seconds
    32°F (0°C)   − 4 to 10 seconds
    95°F (35°C) and above − 1 second

Pull the plug from the thermo-time switch, actuate the starter again when the test lamp should light up for one second and then go out. Finally, connect the test lamp between terminal 'TH' and earth. Actuate the starter when the test lamp should remain illuminated as long as the starter is actuated.

5   Should any of the tests not produce the required results, the components should be renewed.

Fig. 3.48. Crankshaft pulley used in conjunction with fuel injection system

Fig. 3.49. Extracting fuel injection pump drivebelt

Fig. 3.50. Testing start valve (fuel injection system)

Fig. 3.51. Checking thermo-time switch (fuel injection system)

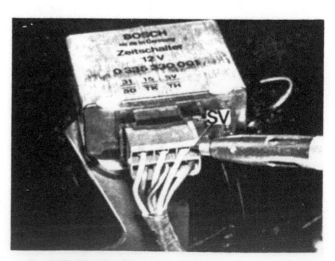

Fig. 3.52. Checking time switch (fuel injection system)

29 Emission control systems - description

Later models are fitted with one or more of the following systems dependent upon the regulations in force within the territories in which the car is to operate. The systems are:

A Crankcase emission control system (see Chapter 1, Section 37).
B Exhaust emission and fuel evaporative control system with air pump.
C Exhaust emission control system without air pump.
D Independent fuel evaporative emission control system.

30 Emission control system (with air pump) - maintenance

1 The correct operation of this system is dependent upon the correct timing of the ignition, as described in Chapter 4, and the correct adjustment of the Solex 32/32 DIDTA carburettor, as described in Section 13 of this Chapter.
2 Check the adjustment of the air pump drivebelt. The tension is correct when the belt can be depressed by between 0.19 and 0.39 in (5.0 and 10.0 mm) at the centre of its top run.
3 Check that all the system hoses are in good condition and their security.

31 Emission control system (without air pump) - maintenance

1 Correct operation of this system is dependent upon the contact breaker points gap being correctly set by a dwell meter and the ignition timing set to Specifications (see Chapter 4). The carburettor must also be correctly adjusted, as described in Section 13.
2 Every 20,000 miles (32,000 km) clean the exhaust gas recirculation pipes. The pipes run between the exhaust and intake manifolds and connect with the diaphragm valve and the cyclone filter.
3 Do not scratch the inside of the pipes and make sure that the pipe unions are re-sealed perfectly.
4 Renew the cyclone filter every 56,000 miles (84,000 km), also the diaphragm valve.

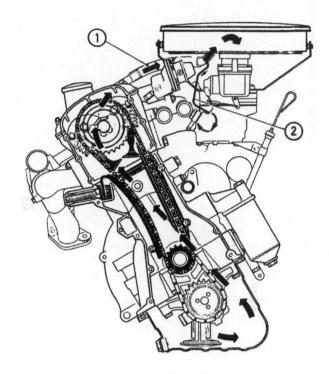

Fig. 3.53. Crankcase ventilation system (1) vent (2) vacuum control

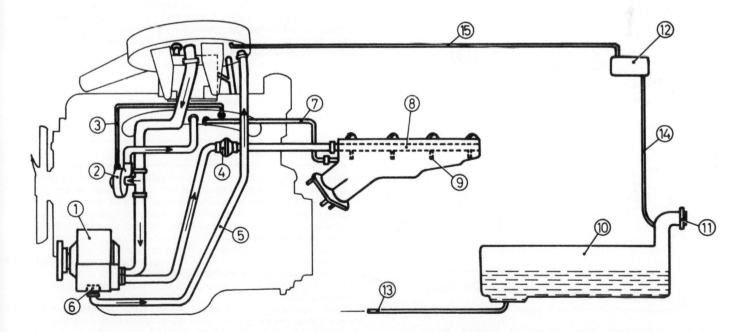

Fig. 3.54. Emission control system (with belt driven air pump)

1 Air pump	5 Air return	8 Air distribution pipe	12 Vapour storage
2 Control valve	6 Pressure regulator	9 Ejectors	13 Fuel supply line
3 Connecting pipe	7 Exhaust gas recirculation pipe	10 Fuel tank	14 Vapour purge line
4 Non-return valve		11 Non-vented cap	15 Extractor line

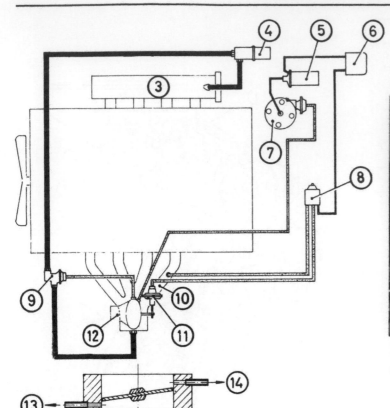

Fig. 3.57. Emission control air pump drivebelt adjustment diagram

A = 0.19 to 0.39 in (5.0 to 10.0 mm)

Fig. 3.55. Exhaust emission control system (without air pump)

3	*Exhaust manifold*	10	*Inlet manifold*
4	*Cyclone filter*	11	*Dashpot*
5	*Ignition coil*	12	*Carburettor*
6	*Speed sensitive relay*	13	*Vacuum pipe to*
7	*Distributor*		*vacuum unit*
8	*Solenoid valve*	14	*Vacuum pipe to*
9	*Diaphragm valve*		*diaphragm valve*

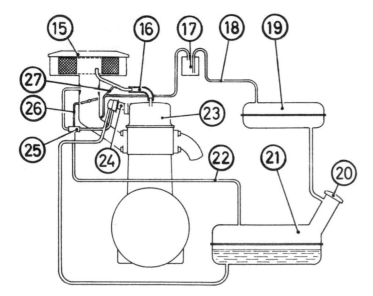

Fig. 3.56. Fuel evaporative control system

15	*Air cleaner*	22	*Fuel expansion pipe*
16	*Crankcase vent*	23	*Rocker cover*
17	*Carbon filter*	24	*Fuel pump*
18	*Vapour purge line*	25	*Fuel return control valve*
19	*Vapour storage*	26	*Vacuum hose*
20	*Non-vented cap*	27	*Crankcase vacuum control*
21	*Fuel tank*		

32 Evaporative emission control system - maintenance

1   This system ensures that fuel fumes from the rear mounted fuel tank are initially collected in a storage tank located under the rear parcels shelf within the luggage boot and then passed through a carbon activated filter which is mounted within the engine compartment on the right-hand bulkhead. When the engine is running, excess fumes are drawn into the engine combustion chambers and burned in the normal way.

2   The system is free from maintenance except for replacement of the screen at the base of the charcoal filter should it become clogged and to occasionally check the security of the system hoses.

33 Manifolds and exhaust system

1   Removal and refitting of the inlet and exhaust manifolds and the air container (fuel injection) is quite straight-forward, but certain models will have coolant hose connections to the inlet manifold for the purpose of induction heating, car interior heater or automatic choke. In these cases, the cooling system will have to be drained first (photo).

2   Note the separate gaskets used at the individual manifold ports and make sure that they are correctly installed to clean, dry flanges.

3   A sectional exhaust system is used and incorporates a primary and secondary silencer. It is recommended that the complete system is removed from under the car whenever a section or silencer is to be renewed (photos).

4   When installing new components, use new flange gaskets and clamps. Do not tighten the clamps fully until the complete system has been installed and its suspended attitude checked for proximity to adjacent suspension and bodywork components.

5   Refer to Fig. 3.63. It is imperative that the exhaust bracket is attached to the rear of the transmission unit in the following way otherwise resonance can occur when the car is running. Loosen retaining plate (1), press the support (2) gently into contact with the exhaust pipe. Tighten the retaining plate bolts to both the transmission and the support. Finally tighten the exhaust pipe clamp nuts (3).

Fig. 3.58. Exhaust gas recirculation pipe connecting with diaphragm valve

Fig. 3.59. Exhaust gas recirculation pipe connecting with cyclone filter

Fig. 3.60. Location of emission control system cyclone filter (1)

Fig. 3.61. Location of evaporative control carbon filter

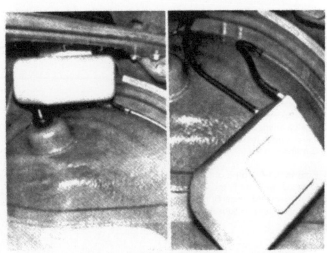

Fig. 3.62. Location of evaporative control vapour storage tank

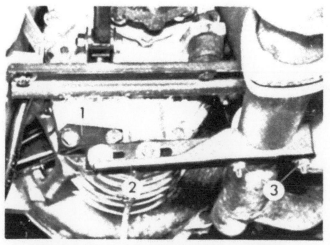

Fig. 3.63. Exhaust mounting at rear of transmission

1   Retaining plate   3   Clamp nuts
2   Support

33.1a Removing inlet manifold with carburettor

33.1b Exhaust manifold

33.3a Exhaust pipe flange connection (exterior)

33.3b Exhaust pipe flange connection (interior)

34 Fault diagnosis - fuel system general

Symptom	Reason/s
Excessive fuel consumption	Air cleaner choked
	Leakage from pump, tank or fuel lines
	Incorrect valve clearances
	Faulty or incorrectly adjusted ignition components
	Tyres under-inflated
	Binding brakes
Insufficient fuel delivery or weak mixture	Fuel tank vent pipe blocked
	Clogged fuel line filter
	Inlet manifold or carburettor gasket flange gaskets leaking
	Fuel pipe unions loose

35 Fault diagnosis - carburettors

Symptom	Reason/s
Excessive fuel consumption	Float chamber flooding (incorrect fuel level or badly seating needle valve or valve body loose in carburettor cover). Mixture too rich
Weak mixture or insufficient fuel delivery	Incorrectly adjusted carburettor Fuel pump lid or pipe connections loose Fuel pump diaphragm split Faulty fuel pump valves Fuel inlet needle valve clogged or stuck

36 Fault diagnosis - fuel injection system

Symptom	Reason/s
Engine will not start from cold but pump operates	Empty fuel tank Ignition fault Start valve faulty Fuel pump pressure too low (may be caused by corroded earth lead contact or worn pump)
Engine will not start when warm but pump operates	Empty fuel tank Start valve not cutting off Fuel pump pressure too high Fuel pump pressure too low
Fuel pump does not operate	Fuse blown Break in supply lead or corroded earth
Erratic idling	Induction (resonator pipes) leaking Throttle valve setting not synchronised with pump Injection pump incorrectly adjusted Throttle valve sticks
Idling speed too high or too low at normal operating temperature	Incorrect slow-running adjustment Incorrect ignition setting Incorrect accelerator linkage adjustment Throttle valve not synchronised with injection pump Fuel injection pump Incorrectly adjusted
Engine backfires on overrun	Throttle valve not synchronised with injection pump Throttle valve not returning fully
Lack of power	Fuel pump pressure too low Throttle valve not synchronised with injection pump Throttle valve not fully opening
Excessive fuel consumption	Leaky start valve Throttle valve not synchronised with injection pump Injection pump incorrectly adjusted

37 Fault diagnosis - emission control system

Symptom	Reason/s
Fumes emitted from engine and condensation in rocker cover	Break in crankcase ventilation system hoses Pre-heater not set to cold position in winter conditions
Fumes emitted from exhaust	Air pump drivebelt slack (if fitted) System hoses loose Incorrect ignition setting Exhaust gas recirculation pipes corroded Cyclone filter clogged Diaphragm valve requires renewal

Chapter 4 Ignition system

Contents

Specifications

System

Early 1500	6 volt negative earth. Coil ignition
All other models	12 volt negative earth. Coil ignition

Firing order ... . 1 - 3 - 4 - 2

Distributor

Type:

All models except 1502, 2002 TI and 2002 TII	Bosch IFUR 4
1502	Bosch JF4D4
2002 TI	Bosch IFR 4
2002 TII	Bosch IFDR 4
Rotational direction	Clockwise
Contact breaker gap	0.016 in. (0.4 mm)

Dwell angle:

1500/1600	59 to 61°
1502	59 to 65°
1602	61 to 65°
2000/2002 except TI	59 to 65°
2002 TI	59 to 61°

Static ignition timing:

All models except 2002 TI and 2002 TII	3° BTDC
2002 TI and 2002 TII	TDC

Dynamic ignition timing:

All models except 1502, 2002 TI and 2002 TII	25° BTDC @ 1400 rpm
1502	25° BTDC @ 1900 rpm
2002 TI	25° BTDC @ 2200 rpm
2002 TII	25° BTDC @ 2400 rpm
2002 with emission control and air pump	25° BTDC @ 2000 rpm

Maximum centrifugal advance:

1500/1600/1602	18°
2002	16°
2002 (automatic transmission)	2 to 6°
2002 TI and TII	16°

Maximum vacuum advance:

All models except 1502	4 to 6°
1502 ... ... :..	10°

Ignition advance (engine at operating temperature, vacuum pipe disconnected):

Rpm							1502	1500/1600/1602	2002
1000	...	...	...	...	...	...	5 to 10°	23 to 27°	21 to 25°
1500	...	...	...	...	...	...	—	25 to 29°	26 to 30°
2000	...	...	...	...	...	...	26 to 30°	30 to 34°	31 to 35°
2500	...	...	...	...	...	...	32 to 37°	34 to 38°	36 to 40°
2700	...	...	...	...	...	...	—	—	38 to 42°
3000	...	...	...	...	...	...	34 to 38°	38 to 42°	—
3500	...	...	...	...	...	...	—	40 to 44°	—
3800	...	...	...	...	...	...	—	42 to 46°	—
4000	...	...	...	...	...	...	38 to 45°	—	—

Rpm							2002 Auto. transmission	2002 TI	2002 TII
1000	...	...	...	...	...	...	11 to 15°	18 to 22°	2 to 7°
1500	...	...	...	...	...	...	17 to 21°	23 to 27°	12 to 17°
2000	...	...	...	...	...	...	23 to 27°	28 to 32°	18 to 22°
2500	...	...	...	...	...	...	30 to 33°	33 to 37°	24 to 28°
2700	...	...	...	...	...	...	—	35 to 39°	
3000	...	...	...	...	...	...	37 to 41°	—	28 to 32°
3500	...	...	...	...	...	...	41 to 45°	—	30 to 34°
3800	...	...	...	...	...	...	42 to 46°	—	—
4000	...	...	...	...	...	...	—	—	—
Vacuum advance starts between	...	...	...	...	...	...	4.72 and 5.91 in Hg. (120.0 and 150.0 mm Hg)		
Vacuum advance ends between	...	...	...	...	...	...	7.68 and 8.27 in Hg. (195.0 and 210.0 mm Hg)		

Coil

Type:									
1500									
Early 6V	...	...	...	...	...	...	...	...	Bosch TE6B4
Later 12V	...	...	...	...	...	...	...	...	Bosch TE12V
1502/1600/1602	...	...	...	...	...	...	...	...	Bosch TE12V
2002	...	...	...	...	...	...	...	...	Bosch KW12V
2002 (automatic transmission)	...	...	...	...	...	...	...	Bosch KW12V	
2002 TI and 2002 TII	...	...	...	...	...	...	...	Bosch K12V	

Condenser

Capacity	...	...	...	...	...	...	..	...	0.23 to 0.32 uf

Spark plugs

Type:			1502	1500/1600/1602	2002	2002 Auto. Transmission	2002 TI	2002 TII
Bosch	...	...	W145T30	W200T30	W200T30 W175T30*	W200T30 W175T30*	W200T30	WG200T30 W175T30*
Beru	...	...	—	200/14/3A	200/14/3A 175/14/3A*	200/14/3A 175/14/3A*	200/14/3A	G200/14/3 175/14/3A*
Champion	...	...	—	N8Y	N8Y N9Y*	N9Y	N8Y	N9Y

* Used with 9.5 : 1 compression ratio and in engines with redesigned combustion chambers marked E12 on cylinder head.

Spark plug gap	...	...	...	...	...	...	...	...	0.024 to 0.028 in (0.6 to 0.7 mm)

Torque wrench settings

									lb/ft	Nm
Spark plugs	...	...	...	...	...	...	...	...	22	30

1 General description

In order that the engine can run correctly it is necessary for an electrical spark to ignite the fuel/air mixture in the combustion chamber at exactly the right moment in relation to engine speed and load. The ignition system is based on feeding low tension (LT) voltage from the battery to the coil where it is converted to high tension (HT) voltage. The high tension voltage is powerful enough to jump the spark plug gap in the cylinders many times a second under high compression pressures, providing that the system is in good condition and that all adjustments are correct.

The ignition system is divided into two circuits. The low tension circuit and the high tension circuit.

The low tension (sometimes known as the primary) circuit consists of the battery lead to the control box, lead to the ignition switch, lead from the ignition switch to the low tension or primary coil windings (terminal +), and the lead from the low tension coil windings (coil terminal —) to the contact breaker points and condenser in the distributor.

The high tension circuit consists of the high tension or secondary coil windings, the heavy ignition lead from the centre of the coil to the centre of the distributor cap, the rotor arm, and the spark plug leads and spark plugs.

The system functions in the following manner. Low tension voltage is changed in the coil into high tension voltage by the opening and closing of the contact breaker points in the low tension circuit. High tension voltage is then fed via the carbon brush in the centre of the distributor cap to the rotor arm of the distributor cap, and each time it comes in line with one of the four metal segments in the cap, which are connected to the spark plug leads, the opening and closing of the

contact breaker points causes the high tension voltage to build up, jump the gap from the rotor arm to the appropriate metal segment and so via the spark plug lead to the spark plug, where it finally jumps the spark plug gap before going to earth.

The ignition is advanced and retarded automatically, to ensure the spark occurs at just the right instant for the particular load at the prevailing engine speed.

The ignition advance is controlled both mechanically and by a vacuum operated system. The mechanical governor mechanism comprises two weights, which move out from the distributor shaft as the engine speed rises due to centrifugal force. As they move outwards they rotate the cam relative to the distributor shaft, and so advance the spark. The weights are held in position by two light springs and it is the tension of the springs which is largely responsible for correct spark advancement.

The vacuum control consists of a diaphragm, one side of which is connected via a small bore tube to the carburettor, and the other side to the contact breaker plate. Depression in the inlet manifold and carburettor, which varies with engine speed and throttle opening, causes the diaphragm to move, so moving the contact breaker plate, and advancing or retarding the spark. A fine degree of control is achieved by a spring in the vacuum assembly.

On cars equipped with an exhaust emission control system, a speed sensitive relay, solenoid valve and carburettor dashpot are used to regulate the operation of the distributor advance vacuum circuit to minimise the emission of fumes during certain operational conditions particularly during deceleration with the accelerator pedal released. Refer to Chapter 3, Fig. 3.33. On all models a resistor is fitted as standard in the coil primary circuit to prevent voltage drop and difficult starting when the starter motor is actuated.

2 Contact breaker - adjustment and lubrication

1   To adjust the contact breaker points to the correct gap, first pull off the two clips securing the distributor cap to the distributor body, and lift away the cap. Clean the cap inside and out with a dry cloth. It is unlikely that the four segments will be badly burned or scored, but if they are the cap will have to be renewed (photo).

2   Inspect the carbon brush in the top of the distributor cap and press it into its recess to test the spring (photo).

3   Check the condition of the rotor arm and renew it if the metal contacts are burned or any cracks are evident.

4   Prise the contact breaker points apart and examine the condition of their faces. If they are rough, pitted, or dirty, it will be necessary to remove them for resurfacing, or for replacement points to be fitted.

5   Presuming the points are satisfactory, or that they have been cleaned and replaced, measure the gap between the points by turning the engine over until the heel of the breaker arm is on the highest point of the cam.

6   The gap between the points should be 0.016 in (0.4 mm). If the gap varies from this, release the securing screw which holds the contact arm to the baseplate and insert a screwdriver between the two 'pips' using them as fulcrum points to move the contact arm as required (photo).

7   When the gap is correct, retighten the securing screw.

8   Apply two drops of engine oil to the felt pad on the top of the distributor shaft and apply a smear of high melting point grease to the high points of the cam.

9   Install the rotor arm and cap.

10 It must be emphasised that adjustment of the contact points by the method just described is not as precise as adjustment of the dwell angle. The dwell angle is the number of degrees through which the distributor cam turns during the period between the closure and opening of the contact breaker points. The angle can only be checked by using a dwell meter. If the angle is larger than that specified, increase the points gap, if too small, reduce the points gap. Setting the dwell angle is essential on cars equipped with exhaust emission control systems.

3 Contact breaker points - removal and refitting

1   Prise off the spring clips and remove the distributor cap and rotor arm.

2   Press the spring type breaker arm in towards the centre of the distributor and disconnect it from the terminal post and the condenser lead tag.

3   Unscrew and remove the screw which secures the fixed breaker arm to the baseplate and then lift the contact breaker assembly upwards.

4   Inspect the faces of the contact points. If they are only lightly burned or pitted then they may be ground square on an oilstone or by rubbing a carborundum strip between them. Where the points are found to be severely burned or pitted, then they must be renewed and at the same time the cause of the erosion of the points established. This is most likely to be due to poor earth connections from the battery negative lead to body earth or the engine to earth strap. Remove the connecting bolts at these points, scrape the surfaces free from rust and corrosion and tighten the bolts using a star type lockwasher. Other screws to check for security are: the baseplate to distributor body securing screws, the distributor body clamp bolt. Looseness in any of these could contribute to a poor earth return.

5   Finally check the condenser, as described in Section 4.

6   Refitting is a reversal of removal, but then adjust the points gap or dwell angle, as described in the preceding Section.

4 Condenser (capacitor) - removal, testing and refitting

1   The condenser ensures that with the contact breaker points open, the sparking between them is not excessive to cause severe pitting. The condenser is fitted in parallel and its failure will automatically cause failure of the ignition system as the points will be prevented from interrupting the low tension circuit.

2   Testing for an unserviceable condenser may be effected by switching on the ignition and separating the contact points by hand. If this action

2.1 Removing distributor cap

2.2 Removing rotor arm

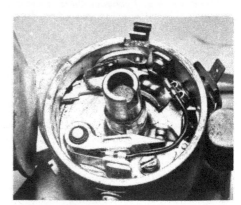
2.6 Contact points

5.3 Removing distributor

5.5 Distributor positioned ready for installation

is accompanied by a blue flash then condenser failure is indicated. Difficult starting, missing of the engine after several miles running or badly pitted points are other indications of a faulty condenser.

3   The surest test is by substitution of a new unit.

4   The condenser is mounted on the outside of the distributor body and is removed by unscrewing its clamp screws and detaching its lead from the contact breaker terminal post.

5   Distributor - removal, installation and ignition timing

1   The distributor is located at the rear of the cylinder head and the cap complete with leads should first be removed.

2   Disconnect the vacuum pipe from the distributor vacuum capsule.

3   Release the clamp bolt on the distributor housing and withdraw the distributor upwards (photo).

4   *On carburettor models* before installing the distributor, apply a spanner to the crankshaft pulley nut and turn the crankshaft until no. 1 piston is at TDC on its compression stroke. This position can be found by turning the crankshaft while holding a finger over no. 1 spark plug hole until pressure can be felt being generated. Continue turning so that the leading notch (static setting) on the crankshaft pulley passes the pointer on the timing chain cover and then align the second notch (TDC) with the pointer.

5   Hold the distributor over its hole in the drive housing so that the vacuum capsule is on the right-hand side and in alignment with the rear corner of the exhaust manifold when looking towards the front of the engine. Set the mark on the end of the rotor arm about 1.4 in (35.6 mm) in an anticlockwise direction from its alignment mark on the rim of the distributor body. Insert the distributor into the drive housing and as the driven gear meshes with the camshaft drive gear, the rotor arm will turn and align with the rim mark. If alignment of the marks is incorrect, withdraw the distributor and turn the rotor arm slightly in either direction as necessary, on a trial and error basis (photo).

6   The crankshaft can now be turned in an anticlockwise direction until the static timing mark is opposite the pointer on the timing chain cover and the distributor turned until the contact points are just opening then tighten the housing clamp bolt. To check the moment of opening of the points, a test bulb may be connected between the contact spring arm and earth and the ignition switched on.

   On some models, the rotor arm incorporates a sliding contact bar. Its purpose is to prevent over revving of the engine and this is achieved by the action of centrifugal force which throws the slide bar out at very high engine speeds and shorts the HT circuit. Uneven running will immediately be detected when this happens and the accelerator should be eased back or a change made to a higher gear.

   For the purposes of ignition timing, the centre of the contact at the end of the rotor arm should be used as the setting mark for alignment with the line on the rim of the distributor body, otherwise ignition timing procedure is exactly as described earlier in this paragraph. Do not confuse the rotor arm and slide bar contacts.

7   A stroboscope should be used to check the ignition timing more precisely. To do this, run the engine to normal operating temperature, disconnect the vacuum pipe from the distributor and plug the pipe. Connect the stroboscope in accordance with the maker's instructions (usually between the end of no. 1 spark plug lead and no. 1 spark plug terminal).

8   On models with more than one notch on the rim of the crankshaft pulley, whiten the static timing notch and the pointer with chalk or paint and with the engine idling, point the stroboscope at the pointer on the timing chain cover. If the timing is correct the two marks will appear stationary and in exact alignment. If they are out of alignment, move the distributor slightly in either direction until the marks coincide.

9   On some models including those with emission control systems, a third notch is located on the crankshaft pulley wheel and this mark can be used employing the stroboscope provided the engine idling speed is set to 2000 rpm.

10   On later models and those with fuel injection systems, the crankshaft pulley has only one (TDC) notch. The distributor can be removed and installed precisely as described previously in this Section, but the timing can only be set by using a stroboscope directed at the steel ball which is inset in the edge of the flywheel. The ball is visible through the inspection hole in the transmission bellhousing. Disconnect the vacuum pipe from the distributor and plug the pipe, connect the stroboscope and adjust the engine idling speed to one of the following settings:

All models except 2002 TI and 2000 TII and those with emission control system with air pump   1400 rpm
2002 TI   2200 rpm
2002 TII   2400 rpm

Models with emission control system with air pump
          2000 rpm

Point the stroboscope at the opening in the bellhousing when half the ball in the flywheel should be showing above the lower edge of the sight hole. If this is not the case, turn the distributor in either direction until the correct alignment of the ball is obtained. Tighten the distributor clamp bolt.

11   Switch off the ignition, remove the stroboscope and reconnect the distributor vacuum pipe.

6   Distributor - overhaul

1   Before dismantling a worn or faulty distributor, make sure that the individual components are available as spares. In the case of a badly worn unit it may be more economical to obtain a new, reconditioned or good secondhand assembly.

2   Remove the distributor, as described in Section 5.

3   Withdraw the cap, rotor arm and contact breaker set.

4   Extract the circlip from the top of the pivot post to which the vacuum capsule link rod is attached.

Fig. 4.1. Rotor arm setting (before installation of distributor

Fig. 4.2. Rotor arm setting (after installation of distributor)

5   Remove the vacuum capsule securing screws, tilt the capsule slightly to disengage the link rod from the pivot and withdraw the capsule.

6   Remove the securing screws and lift the contact breaker plate up and out of the distributor body.

7   Using two screwdrivers inserted under the cam as levers, prise the distributor shaft upwards until the spring retaining ring disengages. Do not extract the felt lubrication pad during this operation or the retaining ring may be lost as it is ejected from the top of the distributor shaft.

8   Secure the distributor in a vice fitted with jaw protectors and drill out the pin which secures the driven gear to the shaft. An 0.118 in (3.0 mm) diameter drill is suitable for this work.

9   Extract the distributor shaft together with the centrifugal advance assembly.

10 If the advance assembly is to be dismantled, mark the components so that they can be refitted in their original locations. If new counterweight springs are installed, the advance curve should be checked out by your dealer or auto-electrician when reassembly is complete.

11 Reassembly is a reversal of dismantling but make sure that the two washers are fitted under the baseplate before the shaft is installed to the body and that the thrust washer is the washer which is furthest from the baseplate.

12 Use a new drive pin in the shaft and check that the shaft top retaining ring is not distorted otherwise use a new one.

13 Install the distributor, as described in the preceding Section.

7   Coil - description and polarity

1   The coil is located within the engine compartment on the right-hand wheel arch.

2   The coil may have an induction transmitter (fuel injection models) plugged into it or a resistor fitted under its upper mounting bolt.

3   HT current should be negative at the spark plug terminals. Make sure this is so by checking that the coil and resistor connections are correctly made otherwise rough idling and misfiring can occur.

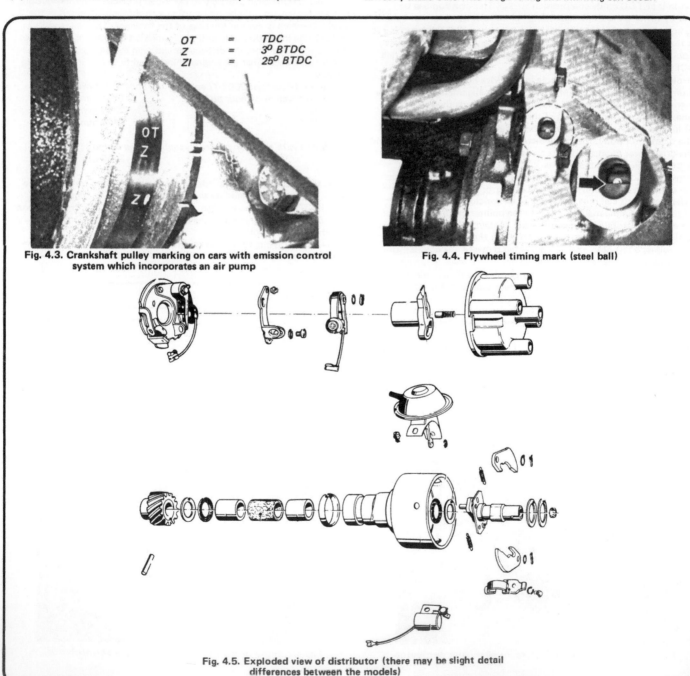

OT	=	TDC
Z	=	3° BTDC
ZI	=	25° BTDC

Fig. 4.3. Crankshaft pulley marking on cars with emission control system which incorporates an air pump

Fig. 4.4. Flywheel timing mark (steel ball)

Fig. 4.5. Exploded view of distributor (there may be slight detail differences between the models)

8 Spark plugs and HT leads

1 The correct functioning of the spark plugs is vital for the correct running and efficiency of the engine. The plugs recommended at the time of publication are as stated in Specifications but always check with the plug manufacturer's guide at time of purchasing new ones as the type of plug specified is sometimes altered in the light of operating experience.

2 At the intervals recommended in 'Routine Maintenance', the spark plugs should be removed, examined, cleaned and if the electrodes are badly eroded or they have been in use for 12,000 miles (19000 km) they should be renewed. The appearance and condition of the spark plugs tells a great deal about the condition of the engine.

3 If the insulator nose of the spark plug is clean and white, with no deposits, this is indicative of a weak mixture, or too hot a plug. (A hot plug transfers heat away from the electrode slowly - a cold plug transfers it away quickly).

4 If the top and insulator nose is covered with hard black looking deposits, then this is indicative that the mixture is too rich. Should the plug be black and oily, then it is likely that the engine is fairly worn, as well as the mixture being too rich.

5 If the insulator nose is covered with light tan to greyish brown deposits, then the mixture is correct and it is likely that the engine is in good condition.

6 If there are any traces of long brown tapering stains on the outside of the white portion of the plug, then the plug will have to be renewed, as this shows that there is a faulty joint between the plug body and the insulator, and compression is being allowed to leak away.

7 Plugs should be cleaned by a sand blasting machine, which will free them from carbon more thoroughly than cleaning by hand. The machine will also test the condition of the plugs under compression. Any plug that fails to spark at the recommended pressure should be renewed.

8 The spark plug gap is of considerable importance, as, if it is too large or too small the size of the spark and its efficiency will be seriously impaired. The spark plug gap should be set to 0.024 in (0.60 mm) for the best results.

9 To set it, measure the gap with a feeler gauge, and then bend open, or close, the outer plug electrode until the correct gap is achieved. The centre electrode should never be bent as this may crack the insulation and cause plug failure, if nothing worse.

10 When replacing the plugs, remember to use new plug washers and replace the leads from the distributor in the correct firing order 1, 3, 4, 2, no. 1 cylinder being the one nearest the radiator.

11 The plug leads require no routine attention other than being kept clean and wiped over regularly.

9 Ignition system - fault diagnosis

Failures of the ignition system will either be due to faults in the HT or LT circuits. Initial checks should be made by observing the security of spark plug terminals, Lucar type terminals, coil and battery connection. More detailed investigation and the explanation and remedial action in respect of symptoms of ignition malfunction are described in the following sub-Sections.

Engine fails to start

1 If the engine fails to start and the car was running normally when it was last used, first check there is fuel in the fuel tank. If the engine turns over normally on the starter motor and the battery is evidently well charged, then the fault may be in either the high or low tension circuits. First check the HT circuit. **Note:** If the battery is known to be fully charged; the ignition light comes on, and the starter motor fails to turn the engine **check the tightness of the leads on the battery terminals** and also the secureness of the earth lead to its **connection to the body.** It is quite common for the leads to have worked loose, even if they look and feel secure. If one of the battery terminal posts gets very hot when trying to work the starter motor this is a sure indication of a faulty connection to that terminal.

2 One of the commonest reasons for bad starting is wet or damp spark plug leads and distributor. Remove the distributor cap. If condensation is visible internally, dry the cap with a rag and also wipe over the leads. Replace the cap.

3 If the engine still fails to start, check that current is reaching the plugs, by disconnecting each plug lead in turn at the spark plug end, and hold the end of the cable about 3/16th inch (5.0 mm) away from the cylinder block. Spin the engine on the starter motor.

4 Sparking between the end of the cable and the block should be fairly strong with a regular blue spark. (Hold the lead with rubber to avoid electric shocks). If current is reaching the plugs, then remove them and clean and regap them. The engine should now start.

5 If there is no spark at the plug leads take off the HT lead from the centre of the distributor cap and hold it to the block as before. Spin the engine on the starter once more. A rapid succession of blue sparks between the end of the lead and the block indicates that the coil is in order and that the distributor cap is cracked, the rotor arm faulty, or the carbon brush in the top of the distributor cap is not making good contact with the spring on the rotor arm. Possibly the points are in bad condition. Clean and reset them as described in this Chapter.

6 If there are no sparks from the end of the lead from the coil, check the connections at the coil end of the lead. If it is in order start checking the low tension circuit.

7 Use a 12v voltmeter or a 12v bulb and two lengths of wire. With the ignition switch on and the points open test between the low tension wire to the coil (it is marked (+) and earth. No reading indicates a break in the supply from the ignition switch. Check the connections at the switch to see if any are loose. Refit them and the engine should run. A reading shows a faulty coil or condenser, or broken lead between the coil and the distributor.

8 Take the condenser wire off the points assembly and with the points open, test between the moving point and earth. If there now is a reading, then the fault is in the condenser. Fit a new one and the fault is cleared.

9 With no reading from the moving point to earth, take a reading between earth and the (−) terminal of the coil. A reading here shows a broken wire which will need to be replaced between the coil and distributor. No reading confirms that the coil has failed and must be replaced, after which the engine will run once more. Remember to refit the condenser wire to the points assembly. For these tests it is sufficient to separate the points with a piece of dry paper while testing with the points open.

Engine misfires

10 If the engine misfires regularly run it at a fast idling speed. Pull off each of the plug caps in turn and listen to the note of the engine. Hold the plug cap in a dry cloth or with a rubber glove as additional protection against a shock from the HT supply.

11 No difference in engine running will be noticed when the lead from the defective circuit is removed. Removing the lead from one of the good cylinders will accentuate the misfire.

12 Remove the plug lead from the end of the defective plug and hold it about 3/16 inch (5.0 mm) away from the block. Restart the engine. If the sparking is fairly strong and regular the fault must lie in the spark plug.

13 The plug may be loose, the insulation may be cracked, or the points may have burnt away giving too wide a gap for the spark to jump. Worse still, one of the points may have broken off. Either renew the plug, or clean it, reset the gap, and then test it.

14 If there is no spark at the end of the plug lead, or if it is weak and intermittent, check the ignition lead from the distributor to the plug. If the insulation is cracked or perished, renew the lead. Check the connections at the distributor cap.

15 If there is still no spark, examine the distributor cap carefully for tracking. This can be recognised by a very thin black line running between two or more electrodes, or between an electrode and some other part of the distributor. These lines are paths which now conduct electricity across the cap thus letting it run to earth. The only answer is a new distributor cap.

16 Apart from the ignition timing being incorrect, other causes of misfiring have already been dealt with under the section dealing with the failure of the engine to start. To recap - these are that:

a) The coil may be faulty giving an intermittent misfire
b) There may be a damaged wire or loose connection in the low tension circuit
c) The condenser may be short circuiting
d) There may be a mechanical fault in the distributor (broken driving spindle or contact breaker spring)

17 If the ignition timing is too far retarded, it should be noted that the engine will tend to overheat, and there will be a quite noticeable drop in power. If the engine is overheating and the power is down, and the ignition timing is correct, then the carburettor should be checked, as it is likely that this is where the fault lies.

For a COLOR version of this spark plug diagnosis page, please see the inside rear cover of this manual

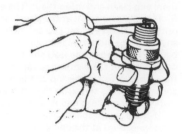

Checking plug gap with feeler gauges

Altering the plug gap. Note use of correct tool

Fig. 4.9a. Spark plug maintenance

White deposits and damaged porcelain insulation indicating overheating

Broken porcelain insulation due to bent central electrode

Electrodes burnt away due to wrong heat value or chronic pre-ignition (pinking)

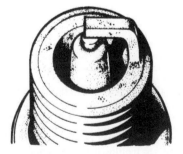

Excessive black deposits caused by over-rich mixture or wrong heat value

Mild white deposits and electrode burnt indicating too weak a fuel mixture

Plug in sound condition with light greyish brown deposits

Fig. 4.9b. Spark plug electrode conditions

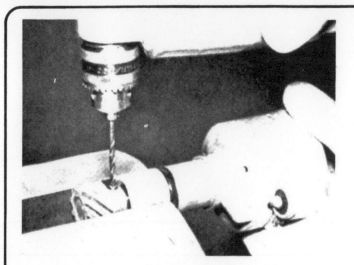

Fig. 4.6. Drilling out distributor gear securing pin

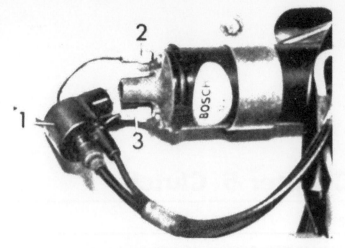

Fig. 4.7. Coil with induction transmitter (1) used on fuel injection models

2 and 3   LT cables to ignition switch and distributor

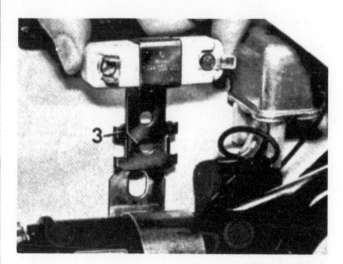

Fig. 4.8. Resistor being fitted to coil mounting bracket (3) lockplate

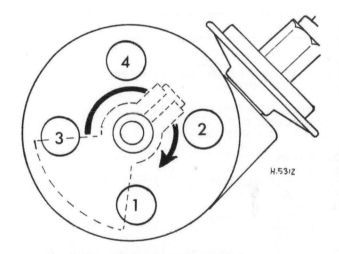

Fig. 4.10. HT lead connecting sequence

Chapter 5 Clutch

Contents

Specifications

Type
1500, 1502, 1600, 1602 and early 2000 and 2002 ... ... Single dry plate coil spring pressure plate
2000 and 2002 (1971 on) ... ... Single dry plate diaphragm spring pressure plate

Actuation
Early 1500, 1600, 1602 ... ... Mechanical
Later 1602 and all 1502, 2000 and 2002 ... ... Hydraulic

Driven plate
Diameter:
    1500, 1502, 1600, 1602 ... ... 7.87 in. (200.0 mm)
    2000, 2002 ... ... 8.98 in. (228.0 mm)
Thickness:
    1500, 1502, 1600, 1602 ... ... 0.398 in. (10.1 mm)
    2000, 2002 ... ... 0.417 in. (10.6 mm)

Clutch clearance at release lever (adjustable pushrod fitted) ... 0.12 in. (3.0 mm)

Free-play at pedal ... ... 0.8 to 1.0 in. (20.0 to 25.0 mm)

Total clutch pedal travel ... ... 6.57 in. (167.0 mm)

Master cylinder bore ... ... 0.75 in. (19.05 mm)

Master cylinder stroke ... ... 1.181 in. (30.0 mm)

Slave cylinder bore ... ... 0.813 in. (20.64 mm)

Slave cylinder stroke ... ... 0.787 in. (20.00 mm)

Torque wrench settings

	lb/ft	Nm
Clutch to flywheel bolts	15	21
Pedal pivot bolt	24	33
Master cylinder bolts	17	23

1 General description

1 The type of clutch and actuating mechanism varies according to car model and date of production. Early 1500, 1600 and 1602 models have a mechanically operated clutch while later 1602 models and all 1502, 2000, 2002 models have a hydraulically operated clutch.
2 The clutch on later 2000 and 2002 models is of diaphragm spring type while all 1500, 1502, 1600 and 1602 models and early (up to 1971) 2000 and 2002 models have a clutch of coil spring type.

3 Major components comprise a pressure plate and cover assembly (incorporating a diaphragm spring or coil springs) a driven plate which is fitted with torsion coil springs to cushion rotational shocks when the drive is taken up.
4 The clutch release bearing is of sealed ball type.
5 *With mechanical operation,* depressing the clutch pedal applies force to the release lever through a linkage of short rods, lever and shafts.
6 *With hydraulic operation,* depressing the clutch pedal moves the piston in the master cylinder forwards, so forcing hydraulic fluid

through the fluid line to the slave cylinder. The piston in the slave cylinder moves forward on the entry of the fluid and actuates the clutch release arm by means of a short pushrod.

7   With either system, the release arm pushes the release bearing forward to bear against the release plate.

8   *If a diaphragm spring type clutch* is fitted, the centre of the spring moves inward and by means of its fulcrum points, the outside moves out to disengage the pressure plate from the clutch driven plate.

9   *If a coil spring type clutch is fitted,* the action is similar to that just described, leverage being applied through the release fingers.

10   When the clutch pedal is release the springs force the pressure plate into contact with the high friction linings on the clutch disc and at the same time pushes the clutch disc a fraction forwards on its splines so engaging the clutch disc with the flywheel. The clutch disc is now firmly sandwiched between the pressure plate and the flywheel so the drive is taken up.

11   Where mechanical operation is employed and with early hydraulic systems, periodic adjustment is required to maintain clutch free-movement but on later hydraulic units, adjustment is automatic due to the design of the slave cylinder. No adjustment nuts are fitted to this later type of pushrod.

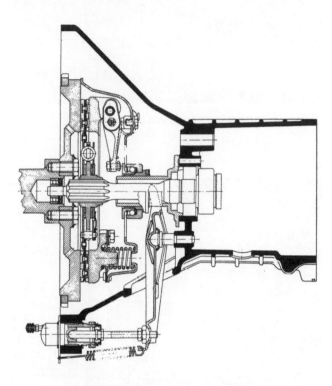

Fig. 5.1. 2000/2002 clutch with coil spring pressure plate

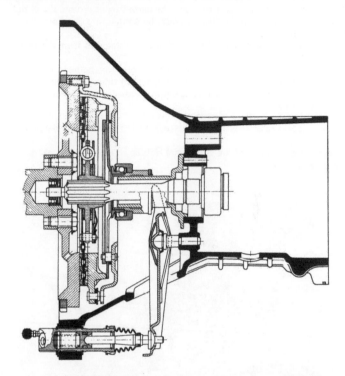

Fig. 5.3. 2000/2002 clutch with diaphragm spring pressure plate

2   Clutch (mechanical operation) - adjustment

1   Release the smaller locknut on the clutch pushrod and detach the return spring from the release arm.

2   Turn the larger adjusting nut until there is a free-movement of 0.12 in. (3.0 mm) between the face of the adjusting nut and the release arm.

3   Tighten the locknut without moving the position of the adjusting nut and reconnect the spring.

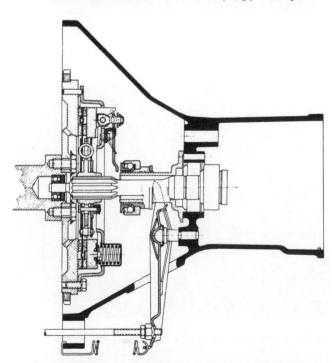

Fig. 5.2. 1500/1502/1600/1602 clutch with coil spring pressure plate

Fig. 5.4. Clutch pushrod (mechanical type)

1   Locknut   2   Adjuster

3   Clutch (early type hydraulic operation) - adjustment

1   The procedure is similar to that just described in the preceding Section.

4   Mechanical linkage and pedal - removal and installation

1   Refer to Fig. 5.6. Working under the car, detach the retaining clip (1) and press the connecting rod (2) from lever on the cross-shaft.
2   Working inside the car, peel back the carpet from around the clutch pedal and extract the clip (3) and press the connecting rod from the clutch pedal.
3   Depress the clutch pedal fully and unscrew the nuts from the tension spring eye bolt. Remove the spring.
4   To remoe the clutch pedal, unscrew and remove the nut from the pivot bolt and having extracted the rubber plug from the pedal support member push the pivot bolt through the member until the clutch pedal can be withdrawn.
5   Refitting is a reversal of removal but grease the pivot and make sure that the spacer washer is installed between the clutch and brake pedal levers.
6   Adjust the length of the spring eye bolt so that the length of the spring is as shown.

Fig. 5.5. Clutch pushrod (hydraulic type)

    1   Locknut   *2   Adjuster*

Fig. 5.7. Clutch (mechanical type) linkage at pedal

    2   Connecting rod   *3   Clip*

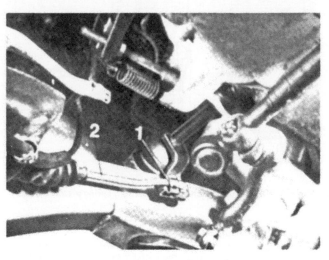

Fig. 5.6. Clutch (mechanical type) linkage

   1   Clip   *2   Connecting rod*

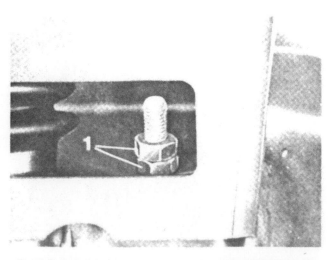

Fig. 5.8. Clutch pedal (mechanical type) spring eyebolt nuts (1)

Fig. 5.9. Clutch pedal (mechanical type) return spring length setting diagram

A = *3.62 in (92.0 mm)*

5 Pedal (hydraulic operation) - removal and installation

1   Working inside the car, peel back the carpet from around the clutch pedal.
2   Disconnect the master cylinder pushrod from the pedal arm by removing pivot bolt, nut, washers and spacers.
3   Unhook the long arm of the pedal pivot return spring from the pedal arm and remove the pedal, as described in paragraph 4 of the preceding Section.
4   Refitting is a reversal of removal but apply grease to all pivots and bushes and note carefully the location of the ends of the pivot spring.
5   Where the clutch slave cylinder is fitted with an adjustable type pushrod, check and adjust the clutch free-movement. Any setting required to systems without an adjustable slave pushrod can be carried out at the master cylinder pushrod.

6 Master cylinder - removal, overhaul and installation

1   On many models, the clutch and brake fluid reservoirs are combined. With this design, syphon out some fluid until it falls to the minimum level and then disconnect the feed pipe to the clutch master cylinder.
2   On later models with independent clutch and brake reservoirs, either syphon out all the fluid from the clutch reservoir or disconnect the feed pipe to the master cylinder and drain the reservoir completely.
3   Disconnect the pipe which runs from the master cylinder to the slave cylinder. Depress the clutch pedal several times to expel fluid from the master cylinder.
4   Working inside the car, disconnect the master cylinder pushrod from the clutch pedal.
5   Unscrew and remove the master cylinder mounting nuts which are accessible within the car and withdraw the unit.
6   Prior to dismantling, clean all dirt from external surfaces.
7   Peel back the rubber boot from the rear end of the master cylinder and extract the circlip.
8   Withdraw the pushrod assembly, piston and spring;
9   Wash all components in methylated spirit or clean hydraulic fluid — nothing else must be used.
10 Examine the surfaces of piston and cylinder bore and if there are scratches, scoring or 'bright' wear areas evident then the master cylinder must be renewed complete.
11 If the components are in good condition, discard all rubber seals and obtain the appropriate repair kit.
12 Install the seals using the fingers only to manipulate them into position and then dip the components in clean hydraulic fluid before installing into the cylinder.
13 Installation is a reversal of removal but bleed the clutch hydraulic system, as described in Section 8.

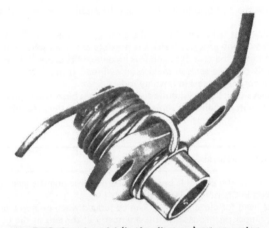

Fig. 5.10. Clutch pedal (hydraulic type) return spring

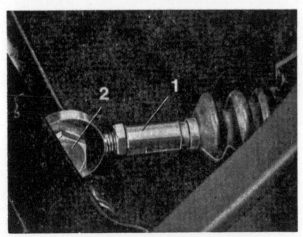

Fig. 5.11. Clutch master cylinder pushrod

1 Sleeve   2 Pedal connecting bolt

Fig. 5.12. Clutch master cylinder securing bolts (1 and 2)

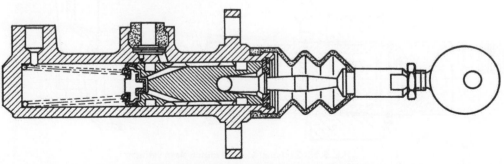

Fig. 5.13. Sectional view of clutch master cylinder

7 Slave cylinder - removal, overhaul and installation

1 Using a pair of pliers, expand the clip which secures the rubber boot to the pushrod end of the slave cylinder. Push the boot and the clip off the cylinder body.
2 Extract the circlip which is on the rear face of the bellhousing mounting flange and withdraw the slave cylinder towards the front of the car.
3 Disconnect the fluid supply pipe from the slave cylinder and plug the pipe. Remove the cylinder (photo).
4 The pushrod and rubber boot can now be detached from the clutch release arm (photo).
5 Overhaul of the cylinder is similar to that described in the preceding Section for the master cylinder.
6 Installation is a reversal of removal but bleed the hydraulic system, as described in Section 8 and where an adjustable pushrod is fitted, check and adjust the clutch free-movement, as described in Section 3.

Fig. 5.14. Clutch slave cylinder mounting

1 Boot clip	2 Circlip

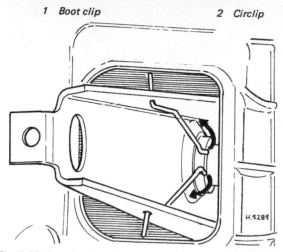

Fig. 5.16. Clutch release bearing attachment to release lever

1   Rubber seal

8 Clutch hydraulic system - bleeding

1 The need for bleeding the cylinders and fluid lines arises when air gets into it. Air gets in whenever a joint or seal leaks or part has to be dismantled. Bleeding is simply the process of venting the air out again.
2 Make sure the reservoir is filled and obtain a piece of 3/16 inch (4.8 mm) bore diameter rubber tube about 2 to 3 feet (0.6 to 0.8 m) long and a clean glass jar. A small quantity of fresh, clean hydraulic fluid is also necessary.
3 Detach the cap (if fitted) on the bleed nipple and surrounding area. Unscrew the nipple ¼ turn and fit the tube over it. Put about ½ inch (12.7 mm) of fluid in the jar and put the other end of the pipe in it. The jar can be placed on the ground under the car.
4 The clutch pedal should then be depressed quickly and release slowly until no more air bubbles come from the pipe. Quick pedal action carries the air along rather than leaving it behind. Keep the reservoir topped-up.
5 When the air bubbles stop, tighten the nipple at the end of a down stroke.
6 Check that the operation of the clutch is satisfactory. Even though there may be no exterior leaks it is possible that the movement of the pushrod from the clutch cylinder is inadequate because fluid is leaking internally past the seals in the master cylinder. If this is the case, it is best to replace all seals in both cylinders.
7 Always use clean hydraulic fluid which has been stored in an airtight container and has remained unshaken for the preceding 24 hours.

9 Clutch - removal

1 Access to the clutch is best obtained by removing the gearbox, as described in Chapter 6, Sections 2 or 3.
2 The need for clutch renewal can be judged when on adjustable type slave cylinder pushrods, the nuts are nearly at the end of the pushrod threads. On pushrods which are not adjustable, renewal of the driven plate is usually required when the overall travel of the end of the release arm is reduced to between 0.67 and 0.75 in. (17.0 to 19.0 mm)
3 Mark the position of the now exposed clutch pressure plate cover in relation to the flywheel on which it is mounted.
4 Unscrew the clutch assembly securing bolts a turn at a time in diametrically opposite sequence until the tension of the diaphragm spring is released. Remove the bolts and lift the pressure plate assembly away (photo).
5 Lift the driven plate (friction disc) from the flywheel.

10 Clutch - inspection and renovation

1 Due to the self-adjusting nature of the clutch it is not always easy to decide when to go to the trouble of removing the gearbox in order to check the wear on the friction lining. The only positive indication that something needs doing is when it starts to slip or when squealing noises on engagement indicate that the friction lining has worn down to the rivets. In such instances it can only be hoped that the friction surfaces on the flywheel and pressure plate have not been badly worn or scored. A clutch will wear according to the way in which it is used. Much intentional slipping of the clutch while driving - rather than the correct

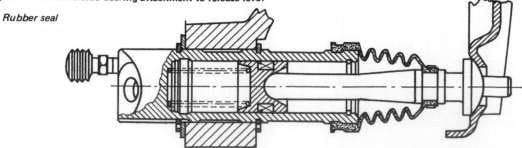

Fig. 5.15. Sectional view of clutch slave cylinder

selection of gears - will accelerate wear. It is best to assume, however, that the friction disc will need renewal at approximately 40,000 mile (64,000 km) intervals at least.

2   Examine the surfaces of the pressure plate and flywheel for signs of scoring. If this is only light it may be left, but if very deep the pressure plate unit will have to be renewed. If the flywheel is deeply scored it should be taken off and advice sought from an engineering firm. Providing it may be machined completely across the face the overall balance of engine and flywheel should not be too severely upset. If renewal of the flywheel is necessary the new one will have to be balanced to match the original.

3   The friction plate lining surfaces should be at least 1/32 in (0.8 mm) above the rivets, otherwise the disc is not worth putting back. If the lining material shows signs of breaking up or black areas where oil contamination has occurred it should also be renewed. If facilities are readily available for obtaining and fitting new friction pads to the existing disc this may be done but the saving is relatively small compared with obtaining a complete new disc assembly which ensures that the shock absorbing springs and the splined hub are renewed also. The same applies to the pressure plate assembly which cannot be readily dismantled and put back together without specialised riveting tools and balancing equipment. An allowance is usually given for exchange units.

11 Release bearing - removal and refitting

1   Whenever the clutch is dismantled for renewal of the driven plate, it is worth renewing the release bearing at the same time.

2   Deterioration of the bearing should be suspected when there are signs of grease leakage or the bearing is very noisy when spun with the fingers.

3   To remove the release arm, prise the spring which secures it to the ball pin pivot over the collar of the ball pin.

4   Withdraw the lever complete with release bearing from the front of the clutch bellhousing (photo).

5   Disconnect the bearing springs and lift it from the release lever.

6   Refitting is a reversal of removal but make sure that the bearing is an exact replacement - several different types have been used. Apply high melting point grease to the ball pivot and the groove inside the bearing hub. Make sure that the conical rubber seal is located between the collar of the ball pivot and the release lever.

12 Clutch - refitting

1   Before refitting the clutch assembly to the flywheel, a guide tool for centralising the driven plate must be obtained. This can either be an old input shaft from a dismantled gearbox or a stepped mandrel made up to fit the centre pilot bearing in the flywheel mounting flange and the inner diameter of the driven plate splines.

2   Examine the condition of the pilot bearing, if it requires renewal, refer to Chapter 1, Section 21.

3   Locate the driven plate against the face of the flywheel, ensuring that the projecting side of the centre splined hub faces towards the gearbox.

4   Offer up the pressure plate assembly to the flywheel aligning the marks made prior to dismantling and insert the retainining bolts finger tight. Where a new pressure plate assembly is being fitted, locate it to the flywheel in a similar relative position to the original by reference to the index marking and dowel positions.

5   Insert the guide tool through the splined hub of the driven plate so that the end of the tool locates in the flywheel spigot bush. This action of the guide tool will centralise the driven plate by causing it to move in a sideways direction (photo).

6   Insert and remove the guide tool two or three times to ensure that the driven plate is fully centralised and then tighten the pressure plate securing bolts a turn at a time and in a diametrically opposite sequence, to the specified torque.

7   Install the gearbox by referring to Chapter 6, and then if an adjustable slave cylinder pushrod is fitted, adjust the clutch free-movement.

7.3 Clutch slave cylinder pipe connection and bleed screw

7.4 Slave cylinder connection to release lever

9.4 Clutch components

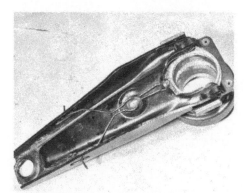

11.4 View of rear face of release lever and bearing

12.5 Centralising the clutch driven plate

13 Fault diagnosis - clutch

Symptom	Reason/s
Judder when taking up drive	Loose engine or gearbox mountings. Badly worn friction surfaces or contaminated with oil. Worn splines on gearbox input shaft or driven plate hub. Worn input shaft pilot bush in flywheel.
Clutch spin (failure to disengage) so that gears cannot be meshed	Incorrect release bearing to spring finger clearance. Driven plate sticking on input shaft splines due to rust. May occur after vehicle standing idle for long period. Damage or misaligned pressure plate assembly.
Clutch slip (increase in engine speed does not result in increase in vehicle road speed - particularly on gradients)	Incorrect release bearing to spring finger clearance. Friction surfaces worn out or oil contaminated.
Noise evident on depressing clutch pedal	Dry, worn or damaged release bearing. Insufficient pedal free travel. Weak or broken pedal return spring. Weak or broken clutch release lever return spring. Excessive play between driven plate hub splines and input shaft splines.
Noise evident as clutch pedal released	Distorted driven plate. Broken or weak driven plate cushion coil springs. Insufficient pedal free travel. Weak or broken clutch pedal return spring. Weak or broken release lever return spring. Distorted or worn input shaft. Release bearing loose on release lever.

Chapter 6 Part 1: Manual gearbox

Contents

Specifications

Application	1500 cc engined models
Type	Long rear extension, floor-mounted gearshift. Four forward speeds (synchromesh) and reverse
Synchromesh type	Porsche
Ratios	
1st	3.816 : 1
2nd	2.07 : 1
3rd	1.33 : 1
4th	1.0 : 1
Reverse	4.153 : 1
Mainshaft endfloat (max.)	0.0241 in. (0.6 mm)
Layshaft endfloat (max.)	0.0079 in. (0.2 mm)
Gear tooth backlash	0.00236 to 0.00591 in. (0.06 to 0.15 mm)
Lubricant capacity	2.2 Imp. pints; 1.25 litres; 2.64 US pints
Application	All models (except 1500 cc engines)
Type (232)	Floor-mounted gearshift. Four forward speeds (synchromesh) and reverse
Synchromesh type	Porsche or Borg-Warner
Ratios	
1st	3.835 : 1
2nd	2.053 : 1
3rd	1.345 : 1
4th	1.0 : 1
Reverse	4.18 : 1
Speedometer drive	2.5 : 1 (teeth 10/4)
Layshaft endfloat	0 to 0.0039 in (0 to 0.1 mm)
Lubricant capacity	1.8 Imp. pints; 1.0 litres; 2.1 US pints
Application	Option on all models (except 1500 cc engines)
Type (235/5)	Floor-mounted gearshift. Five forwards speeds (synchromesh) and reverse

Synchromesh type	Porsche

Ratios

1st	3.368 : 1
2nd	2.16 : 1
3rd	1.58 : 1
4th	1.241 : 1
5th	1.0 : 1
Reverse	4.0 : 1
Speedometer drive	2.5 : 1 (teeth 10/4)

Double gear (end cover) endfloat	0.0039 ± 0.0078 in (0.1 ± 0.2 mm)

Layshaft endfloat	0 to 0.0039 in (0 to 0.1 mm)

Lubricant capacity	2.52 Imp. pints; 1.4 litres; 2.94 US pints

Torque wrench settings

	lb/ft	Nm
Clutch bellhousing to engine bolts (small)	18	25
Clutch bellhousing to engine bolts (large)	34	47
Adaptor plate bolts	15	21
Mainshaft coupling flange nut	108	149
Propeller shaft front coupling nuts to gearbox flange	34	47
Rear cover bolts	18	25
Layshaft end screw (type 235/5)	44	61
Crossmember and rear mounting bolts	18	25

1 General description

1500 models were fitted with a long rear extension four speed gearbox and all other models with a Universal 232 type four speed unit. On later models, a five speed option may be specified and this type of gearbox may be installed as a substitute for the four speed type as described in a later Section of this Chapter. All gearboxes have synchromesh on forward speeds which may be of Porsche or Borg-Warner design.

2 Gearbox (1500 cc engine) - removal and installation

1 Working within the car, pull up the rubber boot from the base of the gearshift lever then draw the foam rubber ring and second boot up the lever.
2 Release the spring retainer, push out the pivot pin and withdraw the gearshift lever.
3 Disconnect the accelerator linkage from the gearbox, also the earth strap.
4 Remove the bolts which secure the starter motor to the clutch bellhousing. Support the starter if necessary by tying it up with a piece of wire.
5 Raise the front of the car until the roadwheels are clear of the ground and support the bodyframe securely on stands. Support the engine sump on a jack.
6 Disconnect the exhaust downpipe from the manifold and the exhaust flexible mounting from the floor pan.
7 Disconnect the exhaust silencer mountings.
8 Disconnect the propeller shaft from the gearbox output shaft and tie the propeller shaft to one side.
9 Disconnect the speedometer cable and reversing lamp switch leads from the gearbox.
10 Disconnect the clutch actuating mechanism according to type, as described in Chapter 5.
11 Release the steering idler (3 bolts) and then turn the steering to full right lock. Make sure that the idler is pushed back towards the engine rear bulkhead. Finally turn the steering back slowly to its central position and rest the idler on the track control arm.
12 Remove the cover plate from the lower half of the front of the clutch bellhousing.
13 Remove the remaining bolts which secure the clutch bellhousing to the engine.
14 Support the gearbox on a jack (preferably trolley type) and unbolt the rear mounting from the gearbox and the body.
15 Lower the gearbox jack carefully together with the engine jack and withdraw the gearbox from below and to the rear of the car. Do not

allow the weight of the gearbox to hang upon the input shaft while it is still engaged with the clutch mechanism. Removal of the gearbox may be eased if the engine is prised forward as far as its flexible mountings will allow but on some models, before this can be done, the air cleaner will have to be removed.
16 Installation is a reversal of removal but observe the following points:
   a) *If a replacement gearbox is fitted to the original engine, remove the two bellhousing to engine hollow positioning dowels. If a replacement engine is fitted to the original gearbox, the dowels must be retained.*
   b) *Tighten all bolts to specified torque.*
   c) *Always install the gearshift lever pivot so that the head of the pivot pin is on the right-hand side when looking towards the front of the car.*
   d) *Refill or top-up the oil level, as appropriate.*
   e) *Adjust the clutch, as described in Chapter 5.*

3 Gearbox (all other cars) - removal and installation

1 Working within the car, pull the dust excluding boot and foam rubber ring up the gearshift lever.
2 Extract the lever ball circlip and retain any shims which are located below it.
3 Working within the engine compartment, unscrew and remove all the upper bolts which secure the clutch bellhousing to the engine.
4 Working under the car, remove the exhaust bracket from the rear of the gearbox, release the bracket clip from the exhaust pipe and swing it aside out of the way (photo).
5 Disconnect the exhaust downpipe from the manifold.
6 Disconnect the front end of the propeller shaft from the output shaft flange of the gearbox, leaving the flexible coupling attached to the propeller shaft.
7 Disconnect the propeller shaft centre bearing from the bodyframe and move the propeller shaft to one side out of the way.
8 Using an Allen key, release the socket screw at the base of the gearshift lever and extract the pivot pin (photo).
9 Disconnect the mechanical or hydraulic clutch actuating mechanism according to type, as described in Chapter 5.
10 Slacken the support bracket bolts and remove the cover plate from the lower part of the front of the clutch bellhousing.
11 Place a wooden support block between the front crossmember and the engine sump.
12 Disconnect the speedometer cable and the leads for the reversing lamp switch from the gearbox (photo).
13 Either raise the front of the car and support securely or place it over a pit.

14 Turn the steering to full right lock and unscrew and remove the remaining bellhousing to engine bolts.
15 Place a jack (preferably trolley type) under the gearbox and support its weight.
16 Unbolt the rear mounting and crossmember from the gearbox and bodyframe (photo).
17 Lower the jack as far as the temporary engine sump block will allow and withdraw the gearbox from below and to the rear of the car. Do not allow the weight of the gearbox to hang upon the input shaft of the gearbox while it is still engaged in the clutch mechanism.
18 Installation is a reversal of removal but observe the following points:
   a) *Tighten all bolts to specified torque.*
   b) *Secure the exhaust pipe bracket to the rear of the gearbox only as described in Chapter 3, Section 33, otherwise resonance may occur during operation of the car.*
   c) *Preload the propeller shaft centre bearing, as described in Chapter 7.*
   d) *Apply grease to the gearshift lever ball before fitting the shims and circlip.*
   e) *Refill or top-up the unit with oil.*
   f) *Adjust the clutch, as described in Chapter 5.*

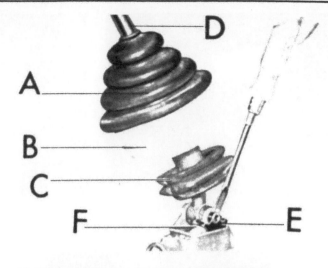

Fig. 6.1. Gearshift lever attachment (1500 cc models)

A Boot
B Foam ring
C Secondary boot
D Lever
E Spring retainer
F Remote control rod

Fig. 6.2. Exhaust silencer mountings (1500 cc)

Fig. 6.3. Speedometer drive cable lock bolt (S) and cable (T) also reversing lamp switch leads (35) on 1500 cc models

Fig. 6.4. Steering idler (1500 cc)

Fig. 6.5. Cover plate screws (1500 cc)

3.4 Exhaust mounting bracket at rear of gearbox

3.8 Releasing gearshift lever socket screw

3.12 Reverse lamp switch and leads

3.16 Gearbox rear mounting crossmember

4 Gearbox (1500 cc engine) - overhaul

1   Drain and discard the oil.

2   Drive out the stop pin which limits the travel of the gearshift remote control rod.

3   Unscrew and remove the remote control rod, plug, spring and plunger from the left-hand side of the gearbox just forward and below the stop pin hole.

4   Pull the gearshift remote control rod slightly to the rear and rotate it in an anticlockwise direction. This will clear the selector finger from the driver.

5   Extract the clutch release lever spring and withdraw the lever and release bearing.

6   From within the clutch bellhousing, remove the bearing cover (three nuts) retaining any shims.

7   Extract the input shaft bearing circlip and shim.

8   Extract the input shaft bearing using a suitable bearing puller.

9   From the rear end of the gearbox, withdraw the sliding coupling shaft.

10 Remove the nuts which secure the rear extension housing to the gearbox casing and then warm the casing around the area of the layshaft bearing to release the bearing from the casing recess. Do not apply excessive heat, remember that the housing and casing are made of light alloy!

11 Using a soft-faced mallet, tap the rear extension housing from the gearbox casing.

12 Withdraw the complete geartrain assembly mounted on the adaptor plate from the gearbox casing.

13 Secure the adaptor plate in the jaws of a vice.

14 Extract the layshaft assembly and shims from the adaptor plate.

15 Withdraw the input shaft and needle roller bearing from the front end of the mainshaft.

16 Drive the reverse gear shaft just enough from the adaptor plate so that the shaft locking ball can be extracted and then drive the shaft in the opposite direction and remove it from the adaptor plate. Withdraw the reverse gearwheel.

17 Set the upper (3rd/4th) selector rod (24) so that fourth gear is engaged and then drive out the fork to rod lock pin.

18 Now push the upper selector rod to the neutral position and set the centre (1st/2nd) selector shaft (27) to engage 2nd gear. Drive out the fork lock pin from the centre rod and then return the rod to the neutral position.

19 Rotate the lower (reverse) selector rod so that the fork is uppermost and drive out the fork lock pin. Withdraw the lower selector rod from the adaptor plate towards the front of the gearbox, catching the released detent balls.

20 Mark 3rd/4th and 1st/2nd selector forks and their rods with a dab of different colour quick-drying paint to identify them and then engage 2nd gear and drive 3rd/4th selector fork from the upper rod. The synchro. sleeve will come with it at the same time.

21 Drive the 1st/2nd selector shaft from the adaptor plate towards the front of the gearbox, and out of its selector fork. Catch the detent balls as they are released.

22 Flatten the tab washer, remove the securing bolts and extract the mainshaft rear bearing retainer.

23 Using a soft-faced mallet, drive the mainshaft from the adaptor plate towards the rear of the gearbox.

24 Clamp the mainshaft in the jaws of a vice fitted with jaw protectors. Extract the circlip from the front end of the mainshaft, withdraw the thrust washer and guide sleeve for 3rd/4th gears.

25 From the front end of the mainshaft, withdraw the 3rd gear and synchro. unit.

26 Extract the spacer and needle bearing, the distance piece and thrust washer.

27 Lift off 2nd gear with synchro. unit, the needle bearing cage, spacer and collar.

28 Remove the sleeve, guide sleeve and 1st gear/synchro. unit.

29 Extract the spacer, collar, needle bearing cage and reverse gearwheel.

30 Mark the relative position of the coupling shaft to the mainshaft and then remove it.

31 Remove the circlip, spring washer, speedometer drive gear and bearing. The last two components will have to be pressed from the shaft. This can be done by supporting the front face of the gear on the open jaws of a vice and driving the mainshaft downwards using a soft-faced mallet.

32 Examine all components for wear and renew as necessary. Renew all oil seals as a metter of routine. Pay particular attention to the Porsche synchromesh units, especially if there has been evidence of noisy gear changing. The teeth on the selector sleeves must have sharp edges and should not be chamfered by wear. Remove the circlip and check that the synchro. ring has an even contact over at least 50% of its periphery. Check that the synchro. locking band is not worn or damaged. 1st gear synchro. can be identified by a key and single locking band. 2nd, 3rd and 4th units have two locking bands. Renew worn components individually.

33 Commence reassembly by fitting the components to the mainshaft in the reverse order to removal. When assembled, the overall length of the geartrain must measure 5.0394 (+0.04 or −0.08 in) or 128.0 (+0.1 or −0.2 mm). In order to obtain this dimension, change the shims which are located in front of the reverse gearwheel.

34 Press the coupling and bearing onto the mainshaft but before pressing on the speedometer drive gear, fit the speedometer drive gear circlip. Now measure the distance between the face of the bearing and the circlip, using a vernier gauge. Record this dimension. Now measure the thickness of the drive gear plus its cup spring. Record this total

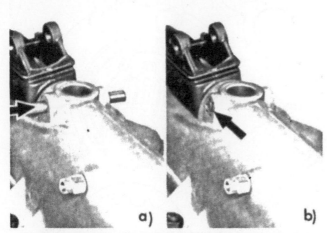

Fig. 6.6. Gearshift remote control rod stop pin (1500 cc)

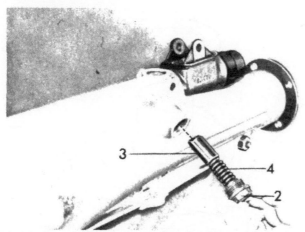

Fig. 6.7. Removing remote control rod plunger assembly (1500 cc)

2 Plug　　　　　　　4 Spring
3 Plunger

Fig. 6.8. Clutch release lever spring (6) release lever (7) collar (B) on 1500 cc gearbox. To remove spring, compress as indicated at 'Y'

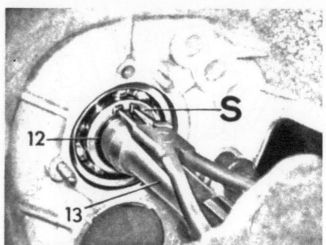

Fig. 6.9. Removing input shaft bearing circlip (1500 cc)

12 Circlip　　　　　　　S Shim
13 Input shaft

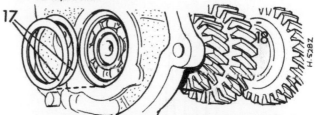

Fig. 6.10. Layshaft installed to adaptor plate (1500 cc)

17 Shims　　　　　　　18 Layshaft

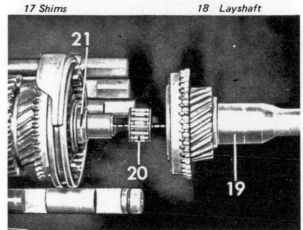

Fig. 6.11. Input shaft (19) needle cage (20) and mainshaft (21) 1500 cc

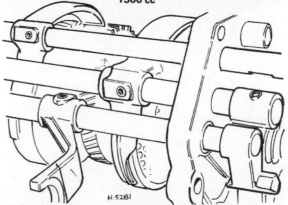

Fig. 6.12. Selector rods and shift forks (1500 cc)

Top 3rd/4th　Centre 1st/2nd　Bottom Reverse

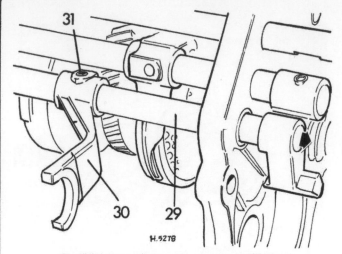

Fig. 6.13. Removing reverse selector rod (1500 cc)

29 Selector rod
30 Shift fork

31 Locking pin

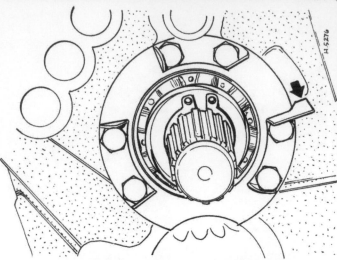

Fig. 6.14. Mainshaft rear bearing retainer and lockplates (1500 cc)

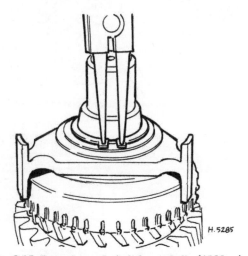

Fig. 6.15. Removing mainshaft front circlip (1500 cc)

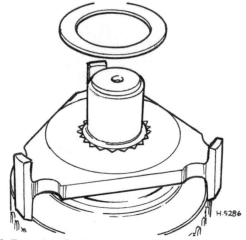

Fig. 6.16. Removing thrust washer and guide sleeve from front end of mainshaft (1500 cc)

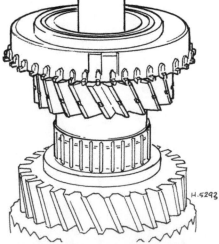

Fig. 6.17. Withdrawing 3rd gear and synchro from front of mainshaft (1500 cc)

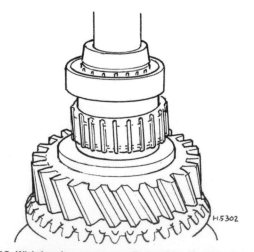

Fig. 6.18. Withdrawing spacer, needle bearing, distance piece and thrust washer from front end of mainshaft (1500 cc)

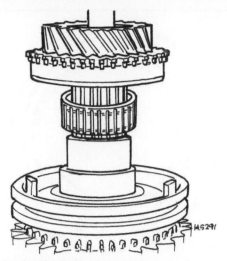

Fig. 6.19. Withdrawing 2nd gear and synchro from front end of mainshaft (1500 cc) together with needle cage and spacers (1500 cc)

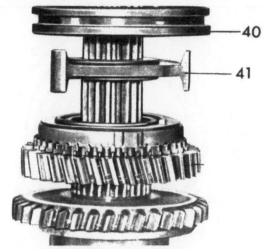

Fig. 6.20. Removing synchro sleeve (40) guide sleeve (41) and 1st gear with synchro (1500 cc)

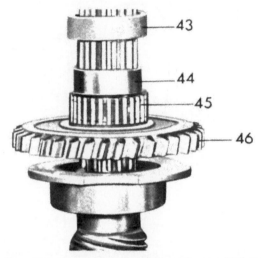

Fig. 6.21. Removing spacer (43) collar, (44) needle cage, (45) and reverse gear (46) from mainshaft (1500 cc)

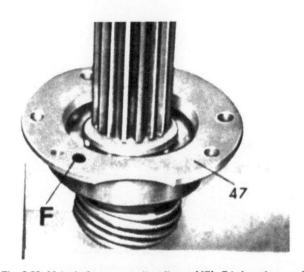

Fig. 6.22. Mainshaft rear coupling flange (47). F is locating mark

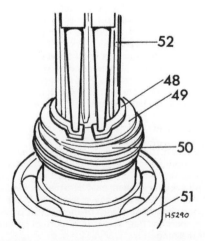

Fig. 6.23. Mainshaft rear circlip (48) spring washer (49) speedo drive gear (50) mainshaft rear bearing (51) and mainshaft (52)

Fig. 6.24. Dismantling Porsche type synchro unit (1500 cc)

53 Circlip   54 Synchro ring

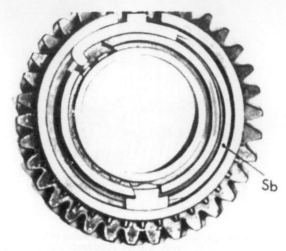

Fig. 6.25. Synchro locking band Sb on 1500 cc first gear

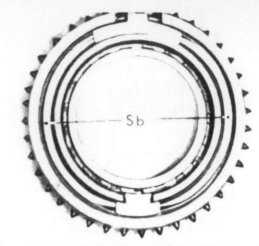

Fig. 6.26. Dual locking bands Sb on 1500 cc 2nd, 3rd and 4th gear synchro units

Fig. 6.27. Measuring overall length of assembled mainshaft gears (1500 cc)

60 Reverse gear   61 Guide sleeve

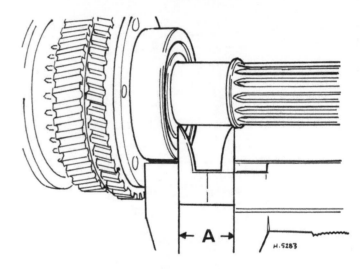

Fig. 6.28. Measuring distance from mainshaft rear bearing to speedo drive gear circlip (1500 cc)

dimension. If the two dimensions are now subtracted the difference will equal the thickness of the shim which must be installed next to the bearing before the speedometer drive gear is pressed onto the mainshaft and the cup spring and circlip finally installed.

35 Before installing the mainshaft to the adaptor plate, install the two detent springs to the adaptor plate.

36 Install the assembled mainshaft to the adaptor plate, fit the mainshaft rear bearing retainer making sure that the lockplate is correctly positioned, otherwise the deflector will not supply lubricant to the speedometer drive gear.

37 Hold the adaptor plate in the vice so that the 3rd/4th detent ball can be located on the end of the detent spring. Insert 3rd/4th selector rod into the adaptor plate so that the cut-out in the rod faces upwards. Push the rod in until the detent ball engages in its rod groove and then rotate the rod into its correct attitude.

38 Insert the interlock ball which locates between 3rd/4th and 1st/2nd selector rods. Place the 1st/2nd detent ball on top of its spring and then press the 2nd gear synchro. sleeve into its 2nd gear engaged position. Install the 1st/2nd selector rod to the adaptor plate, making sure that the detent ball does not engage in the cut-out portion of the rod. Engage the shift fork with the groove in the synchro. sleeve and then slide the selector rod through the fork.

39 Engage the 3rd/4th shift fork with the groove in the 3rd/4th synchro. sleeve and push both components simultaneously into engagement with the 3rd/4th selector rod and hub on the mainshaft.

40 Move 1st/2nd selector rod back to its neutral position.

41 Insert the interlock ball into the adaptor plate so that it lies between the 1st/2nd and reverse selector rods.

42 Locate the detent ball on the reverse detent spring (this has remained in the adaptor plate during overhaul and can only be removed if the sealing plug is first extracted). Install the reverse selector rod to the adaptor plate and fit the reverse shift fork to it.

43 Install new locking tension pins to all the shift forks making sure that the slits in the pins are in alignment with the selector rods.

44 Engage reverse gear with reverse shift fork, pass the shaft through the gear into the adaptor plate so that the locking ball can be inserted and then tap the reverse shaft towards the front of the gearbox to lock it in position.

45 Install the input shaft and needle bearing to the front of the mainshaft.

46 Install the layshaft to the adaptor plate. This is achieved by holding the layshaft assembly in position and driving the layshaft front bearing into position simultaneously onto the layshaft and into the adaptor plate. Shims and packing washers must be fitted to the front face of the layshaft front bearing to take up any space between the recessed face of the bearing and the mating flange of the adaptor plate. This also takes up any layshaft endfloat and ensures correct layshaft to mainshaft gear meshing.

47 Set the gear train in the neutral mode and then position a new gasket on the face of the rear extension housing. Install the adaptor plate gear train assembly to the extension housing.

48 To the front of the gearbox casing, install the bearing circlip, a

0.0394 in (1.0 mm) thick shim and the input shaft bearing.

49 Install a new gasket to the adaptor plate mating face of the gearbox casing and using a depth gauge measure the distance between the surface of the gasket and the inner track of the input shaft bearing. Record this dimension (A) and observe the marking on the input shaft and use the following table to determine the thickness of the input shaft compensating shim.

Depth A	Marking on input shaft	Compensating shim thickness
6.452 in (163.9 mm)	50	0.023 in (0.6 mm)
	40	0.027 in (0.7 mm)
	30	0.031 in (0.8 mm)
6.448 in (163.8 mm)	50	0.019 in (0.5 mm)
	40	0.023 in (0.6 mm)
	30	0.027 in (0.7 mm)
6.444 in (163.7 mm)	50	0.015 in (0.4 mm)
	40	0.019 in (0.5 mm)
	30	0.023 in (0.6 mm)
6.440 in (163.6 mm)	50	0.011 in (0.3 mm)
	40	0.015 in (0.4 mm)
	30	0.019 in (0.5 mm)

50 Now measure the depth in a similar manner from the face of the gasket to the layshaft bearing ring. Record this dimension (D).

51 Measure the length of the layshaft assembly as shown and record

this dimension (E). By subtracting one measurement from the other and in turn subtracting the result from the specified clearance of 0.0079 in (0.2 mm) the shim thickness required can be found.

52 Place the shim on the face of the layshaft rear bearing, heat the gearbox casing as specified for removal and fit the casing to the adaptor plate at the same time forcing the input shaft into its bearing and the layshaft bearing into the casing.

53 Tighten the extension housing/gearbox casing through-bolts to the specified torque.

54 Measure the thickness of the input shaft bearing front circlip and record this dimension (A) before installing the circlip into its groove.

55 Measure the distance between the front face of the bearing inner track and the front face of the circlip, now installed. Record this dimension (B). Subtract one measurement from the other and the difference will be the thickness of the shim which must be installed to the front face of the bearing. To do this, the circlip will again have to be extracted.

56 Measure and record the depth which the front face of the outer track of the input shaft bearing is below the surface of the gear casing. Fit a new gasket to the bearing cover and measure the height of the collar of the cover from the surface of the gasket. By subtracting one measurement from the other, the thickness of the shims which must be installed to the front face of the input shaft bearing can be determined. Fit the selected shim, the bearing cover and gasket and the clutch components.

57 Reconnect the gearshift remote control rod and stake the stop pin. Refit the control rod plunger, spring and screw plug.

58 Refill the gearbox with oil.

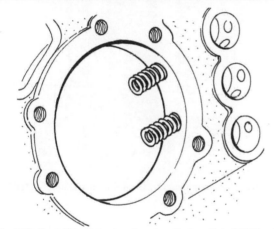

Fig. 6.29. Installing detent springs in adaptor plate (1500 cc)

Fig. 6.30. Installing 3rd/4th selector rod to adaptor plate (1500 cc)

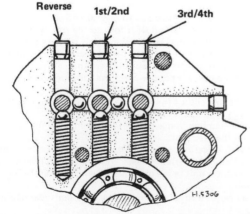

Fig. 6.31. Location of selector rod detent and interlock balls (1500 cc)

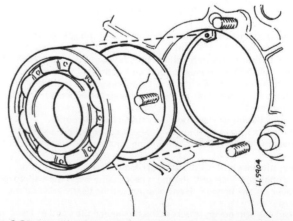

Fig. 6.32. Installing input shaft bearing, shim and circlip (1500 cc)

Fig. 6.33. Measuring distance (A) from input shaft bearing to gear casing flange (1500 cc)

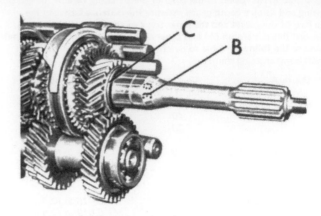

Fig. 6.34. Input shaft number (B) and compensating shim (C)

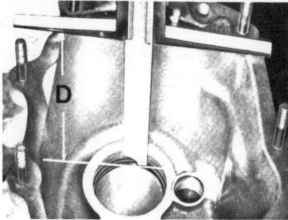

Fig. 6.35. Measuring distance (D) from layshaft bearing to gear casing flange (1500 cc)

Fig. 6.36. Measuring length (E) of layshaft assembly (1500 cc)

5 Gearbox (4 speed type 232) - overhaul

1   Unbolt and remove the bracket and stay from the rear of the gearbox (photo).

2   Push back the spring sleeve on the gearshift remote control rod to expose the coupling pin. Drive out the pin and separate the coupling (photos).

3   Remove the exhaust support bracket from the gear casing.

4   Drain the oil from the unit.

5   From within the clutch bellhousing, remove the clutch withdrawal lever, the release bearing and unbolt and withdraw the input shaft bearing retainer cover. Extract the shim and gasket (photos).

6   Extract the input shaft bearing circlip and shim located behind the circlip (photos).

7   Extract the input shaft bearing using a bearing puller.

8   Unscrew and remove the gearbox end cover bolts.

9   Drive out the two cover positioning pins towards the front of the gearbox.

10 Heat the area around the layshaft front bearing (within the bellhousing) and then pull the main gear casing from the end cover. The cover on the layshaft front bearing will be destroyed and must be renewed.

11 From the end cover, unscrew the hexagonal plug and extract the spring and plunger (gearchange rod) (photo).

12 Move the 3rd/4th selector rod to engage 4th gear and then drive out the pin which secures the 3rd/4th shift fork to the selector rod. Note that the synchro dog teeth must be set in such a way below the pin

that the cut-out in the teeth can receive the pin as it is driven out.

13 Pull the 3rd/4th selector rod towards the front of the gearbox and remove it, catching the detent balls which will be ejected.

14 Rotate the gear selector rod so that the shift dog is uppermost and then withdraw the rod towards the front of the gearbox.

15 Return the 4th gear synchro. sleeve to neutral and withdraw the 3rd/4th shift fork.

16 Push the 1st/2nd selector rod to engage 2nd gear and drive out the pin which retains the 1st/2nd shift fork to the rod, again setting the synchro. dog teeth so that the tooth cut-out receives the pin. Pull the selector rod out towards the front of the gearbox and remove the shift fork. Return 2nd gear to the neutral position.

17 Remove the lock bolt and extract the speedometer drive gear, pinion and bush.

18 From the rear end of the mainshaft, withdraw the flexible buffer and prise out the lockplate.

19 Hold the coupling flange quite still and unscrew the nut. Pull the flange from the mainshaft.

20 Remove the bearing retainer (5 bolts) and extract the shim.

21 Place a thin metal strip between 2nd and 3rd gearwheels to prevent 3rd gear synchro. unit being displaced when the mainshaft rear bearing is removed.

22 With a bearing extractor, remove the mainshaft rear bearing and shim.

23 Withdraw the combined input/mainshaft assembly from the gear casing, inclining the front end from the layshaft as it is withdrawn.

24 Drive the layshaft assembly from the end cover towards the front of the gearbox. Retain the bearing shim.

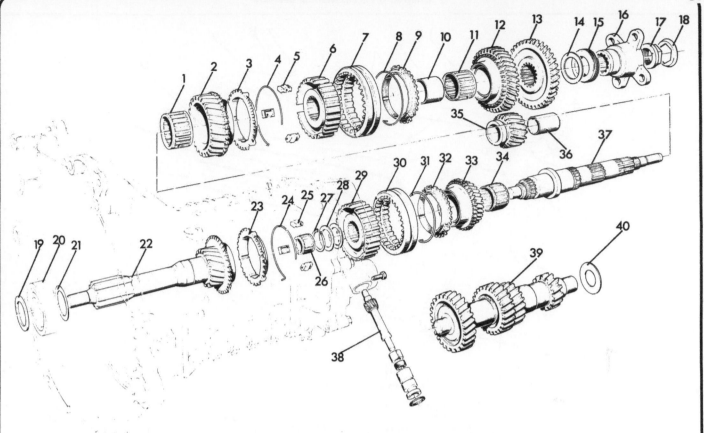

Fig. 6.37. Geartrain components (type 232 four speed) with Borg-Warner synchromesh

1	Needle roller bearing	11	Needle roller bearing	21	Shim	31	Spring

1 Needle roller bearing
2 2nd gear
3 Synchro ring
4 Synchro spring
5 Key
6 Synchro hub
7 Synchro sleeve
8 Spring
9 Synchro ring
10 Sleeve

11 Needle roller bearing
12 1st gear
13 Reverse gear
14 Washer
15 Speedo drive gear
16 Coupling
17 Coupling nut
18 Lockplate
19 Shim
20 Input shaft bearing

21 Shim
22 Input shaft
23 Synchro ring
24 Synchro spring
25 Key
26 Needle bearing
27 Circlip
28 Shims
29 Synchro hub
30 Synchro sleeve

31 Spring
32 Synchro spring
33 3rd gear
34 Needle roller bearing
35 Reverse idler gear
36 Reverse gear bush
37 Mainshaft
38 Speedo driven gear and pinion
39 Layshaft
40 Shim

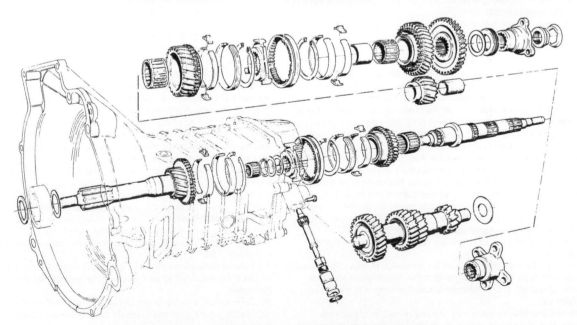

Fig. 6.38. Geartrain components (type 232 four speed) with Porsche synchromesh. Key as Fig. 6.37 except for synchro components

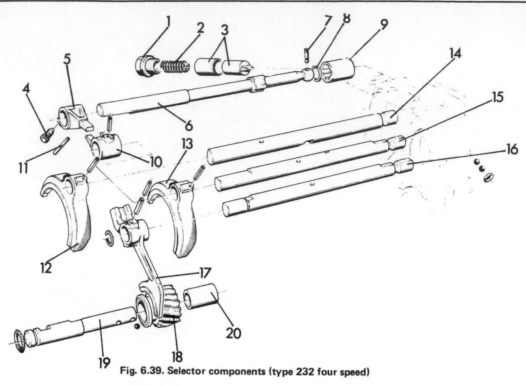

Fig. 6.39. Selector components (type 232 four speed)

1	Threaded plug	6	Gearchange rod	11	Locking pin
2	Spring	7	Locking pin	12	3rd/4th shift fork
I3	Plunger	8	Circlip	13	1st/2nd shift fork
4	Lock bolt	9	Taper bush	14	3rd/4th selector rod
5	Dog	10	Dog	15	1st/2nd selector rod

16	Reverse selector rod
17	Reverse shift fork
18	Reverse idler gear
19	Reverse idler pinion
20	Reverse idler pinion bush

25 From the end cover, drive out the reverse gear pinion together with the reverse selector rod and shift fork. Catch the detent and locking balls.

26 Pull the input shaft together with needle bearing cage from the front end of the mainshaft.

27 Pull 3rd/4th synchro. sleeve from the front of the mainshaft.

28 Extract the circlip, thrust washer and 3rd/4th synchro. components followed by the other mainshaft gears and synchro. units from which the mainshaft will have to be pressed. Porsche or Borg-Warner synchro. units may be encountered and the difference in detail and fitting should be noted from the illustrations.

29 With the gearbox dismantled, check all components for wear and renew as necessary. Renew oil seals as a matter of routine (front and rear bearing covers, speedo. pinion, gearchange rod end cover seal). Where Porsche synchro. units are fitted, check them as described in the preceding Section (paragraph 32). The mainshaft on later cars has been modified to avoid the rapid wear which occurred in the splines of the output end. If a new modified mainshaft is fitted having finer splines, then the coupling flange and rear oil seal must also be renewed. If Borg-Warner synchromesh units are installed and they have become worn, check them in the following way: Fit the synchro. ring to the cone of the gear and using feeler blades, check the clearance. If it is less than 0.032 in (0.8 mm) renew the ring. Press the hub from the sleeve and examine the teeth for wear. The teeth are recessed in the sleeves for engagement with 1st, 2nd and 3rd gears. Engage the springs as shown so that they run in opposite directions on either side of the hub. Engage the sliding keys with the springs and re-install the hubs to the sleeves. It should be noted that when Borg-Warner units were introduced, the meshing angle of the reverse gearwheels was changed from 20° to 15°, the latter being identified by a thin groove round the outer periphery of the gears (photos).

30 If the layshaft end cover bearing is to be renewed, it must be changed from the earlier ball type to the later roller type. Worn gears must be renewed as pairs (layshaft and mainshaft). Layshaft gears are secured with circlips (photo).

31 Commence reassembly by installing the mainshaft components.

32 Oil the front end of the mainshaft, push on 3rd gear needle bearings and 3rd gearwheel (photo).

33 Also to the front of the mainshaft install the 3rd/4th synchro. unit, the wave washer and thrust washer and secure with a circlip (photos).

34 Install 2nd gear needle roller bearing to the rear of the mainshaft (photo).

35 Install 2nd gearwheel (photo).

36 Install 1st/2nd synchro. unit (photo).

37 Install the sleeve and needle roller bearing for 1st gear (photo).

38 Install 1st gear and its synchro. cone.

39 Install reverse gear so that the lead edges of its teeth are towards 1st gear (photo).

40 Place the shim against the hub of the reverse gear.

41 *On transmissions with Porsche type synchro. units,* measure the length of the mainshaft gear train as shown and install shims if necessary (placed on the rear face of the reverse gear hub) to increase the overall length to 5.433 (± 0.039 in) - 138.0 (± 0.1 mm).

42 *With all transmissions,* measure the thickness of the speedometer drive gear and record this dimension (B).

43 Press the mainshaft bearing into the rear cover and then locate a new flange gasket on the cover mating face. Measure the distance between the face of the gasket and the face of the inner track of the bearing. Record this dimension (C). To a theoretical dimension of 5.433 in (138.0 mm) add measurement (B) and then subtract (C). Subtract the result from a theoretical 4.567 in (116.0 mm) and this will be the thickness of the shim required to be placed on the front face of the mainshaft rear bearing.

44 Press the speedometer drive gear onto the mainshaft. It is best to heat the gear first and tap it into position (photo).

45 Measure the depth of the layshaft bearing circlip from the mating flange of the main gearbox casing. Record this dimension (A).

46 Temporarily install the layshaft assembly into the end cover bearing and measure the distance between the end of the layshaft assembly and the face of the gasket laid on the end cover flange. Record this dimension (B).

47 Subtract the dimension (B) from (A) and the result is the thickness of the shim required to be placed in the layshaft bearing recess in the rear cover.

48 Sometimes, a noisy gearbox can be the result of the reverse idler gear teeth rubbing on the layshaft thrust washer. Grind carefully to clear.

49 When the reverse idler gear shaft must be renewed, heat the gearbox end cover to 248°F (120°C) and install it so that the hole in the shaft is towards the centre of the layshaft. The reverse shaft must project

5.1 Bracket and stay at rear of gearbox

5.2a Gearshift remote control rod spring sleeve

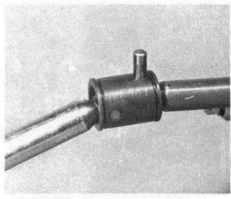
5.2b Gearshift remote control rod joint pin

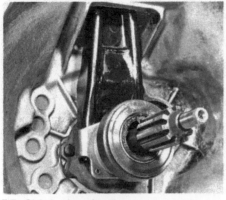
5.5a Clutch release bearing

5.5b Input shaft bearing retainer

5.6a Input shaft bearing circlip

5.6b Input shaft bearing shim

5.11 Removing gearchange rod plunger components

5.29a Rear bearing cover and oil seal

5.29b Old and modified (left) mainshaft and coupling flange

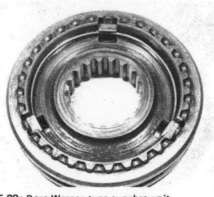
5.29c Borg-Warner type synchro unit

5.30 Interior of gearbox end cover

Fig. 6.40. Gearchange rod oil seal (type 232)

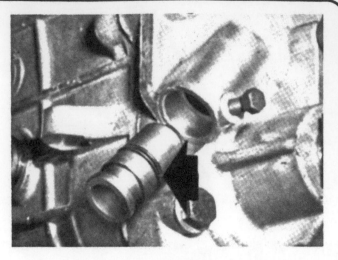

Fig. 6.41. Speedometer pinion oil seal (type 232)

Fig. 6.42. Checking synchro ring clearance (Borg-Warner)

Fig. 6.43. Correct spring engagement (Borg-Warner synchro)

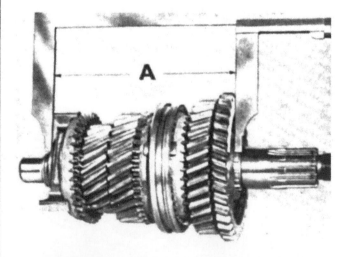

Fig. 6.44. Measuring mainshaft geartrain length (A). Type 232 with Porsche synchromesh

Fig. 6.45. Location of geartrain length adjusting shim - Type 232 with Porsche synchromesh

5.32 Fitting needle bearings and 3rd gear to mainshaft

5.33a Fitting 3rd/4th synchro to mainshaft

5.33b Installing wave washer, thrust washer and circlip to front face of synchro unit

5.33c Installing circlip to front face of 3rd/4th synchro

5.34 Fitting needle bearings to mainshaft

5.35 Installing 2nd gear to mainshaft

5.36 Installing 1st/2nd synchro to mainshaft

5.37 Fitting 1st gear sleeve needle bearings, gear with synchro cone to mainshaft

5.39 Installing reverse gear to mainshaft

5.44 Installing shim and speedo drive gear to mainshaft

5.52a Installing reverse selector rod and reverse idler gear

5.52b Selector rod detent ball caps on end cover

Fig. 6.46. Measuring depth (C) of mainshaft bearing from end cover flange (Type 232)

Fig. 6.47. Measuring depth (A) of layshaft bearing circlip from main gear casing flange (Type 232)

Fig. 6.48. Measuring length (B) of layshaft assembly (Type 232)

0.079 in (2.0 mm) beyond the end of its hole when finally installed.

50 If the reverse idler gear bush must be renewed then the new one must be reamed after pressing in to 0.838 (+ 0.0028) in - 21.3 (+ 0.073) mm.

51 Unscrew the reverse lamp switch and extract the sealing cap adjacent to it, from the end cover.

52 Insert the detent ball into the open reverse lamp switch hole, depress the ball and insert the reverse selector rod and reverse idler gear simultaneously, pushing in the rod until the detent ball engages in the groove in the rod. An alternative method of inserting the detent balls is to remove the three caps from the end cover. If the detent springs have been removed, the shorter one is reverse. With all selector rods, the detent grooves which go only part way round the rods must be positioned towards the detent balls (photos).

53 Locate the previously determined shim (paragraphs 35, 36 and 37) in the layshaft bearing recess in the end cover and press in the layshaft assembly to its bearing.

54 Install the mainshaft assembly complete with input shaft attached to its front end into the end cover, meshing it simultaneously with the layshaft gears (photo).

55 Locate the previously determined shim (paragraphs 32 and 33) on the rear face of the speedometer drive gear and then drive the mainshaft rear bearing onto the mainshaft and into the end cover. Make sure that the groove in the outer track is nearer the rear of the mainshaft (photo).

56 At this stage, re-check the gear tooth engagement between mainshaft and layshaft assemblies. If necessary, this can be improved by altering the thickness of the layshaft bearing shim located on the front face of the layshaft front bearing.

57 Measure the amount by which the mainshaft rear bearing outer track is below the rear cover flange. Record this dimension (A).

58 Place a new gasket on the bearing retainer flange and measure the height of the collar from the face of the gasket. Record this dimension (B).

59 Subtract dimension 'B' from 'A' and the result is the shim thickness required for insertion under the bearing retainer. Bolt on the retainer with the shim and gasket (photos).

60 To the rear end of the mainshaft, fit the lockplate, the coupling flange and nut, tightening to specified torque. Push on the flexible buffer (photo).

61 Engage 1st/2nd shift fork into the groove of the synchro. sleeve.

62 Insert the interlock ball and the detent ball into the hole in the end cover from which the plug was prised.

63 Install the 1st/2nd selector rod, keeping the detent ball depressed and passing the rod through the shift fork.

64 Engage the 3rd/4th shift fork with the groove in the synchro. sleeve. Install the gearchange rod (flat on rod at bottom with dog to right-hand side) making sure that the groove in the splined bush in the end cover is visible through plunger hole before engaging the gearchange rod splines with the bush (photo).

65 Insert the interlock and detent balls into the 3rd/4th selector rod hole. Depress the detent ball and insert the selector rod. Secure the forks to the rods by driving in the locking pins (photos).

66 Install the reversing lamp switch and sealing cap to the ends of the reverse and 1st/2nd selector rods respectively.

67 Install the speedometer pinion and gearchange rod plunger components to the end cover (photo).

68 Now measure the depth of the input bearing inner track below the mating flange of the main gearbox casing. Record this dimension (A).

69 Note the marking (B) on the input shaft. Working from the following table, the thickness of the shim (X) which must be installed to the input shaft (on the front face of top gear) can be established.

Depth (A)	Marking on input shaft (B)	Input shaft shim required (X)
6.059 in (153.9 mm)	*45 to 50*	*0.0196 in (0.5 mm)*
	35 to 40	*0.0236 in (0.6 mm)*
	25 to 30	*0.0276 in (0.7 mm)*
6.055 in (153.8 mm)	*45 to 50*	*0.0157 in (0.4 mm)*
	35 to 40	*0.0196 in (0.5 mm)*
	25 to 30	*0.0236 in (0.6 mm)*
6.051 in (153.7 mm)	*45 to 50*	*0.0118 in (0.3 mm)*
	35 to 40	*0.0157 in (0.4 mm)*
	25 to 30	*0.0196 in (0.5 mm)*

6.047 in (153.6 mm)	45 to 50	0.0078 in (0.2 mm)
	35 to 40	0.0118 in (0.3 mm)
	25 to 30	0.0157 in (0.4 mm)

70 Install the shim to the input shaft. Using grease stick the shim which may have been selected to improve gear mesh (see paragraph 46) to the front face of the layshaft front bearing.

71 Locate a new gasket on the gearbox casing flange and lower the casing over the gear train. During this operation, the input shaft bearing will have to be tapped into position down the input shaft, having first installed an 0.039 in (1.0 mm) thick shim in the bearing recess (photo).

72 Install a new layshaft bearing blanking plug within the bellhousing (photo).

73 Measure the thickness of the input shaft bearing circlip and then install the circlip. Measure the distance which the front face of the circlip is proud of the front face of the input shaft bearing and from this dimension subtract the thickness of the circlip measured previously. The result is the thickness of the shim which must be installed to the front face of the input shaft bearing inner track. To install the selected shim, the circlip must be extracted temporarily.

74 Measure the distance that the outer track of the input shaft bearing outer track is below the rim of the gearcasing (within the bellhousing) and then measure the height of the rim of the collar on the bearing retainer. Subtract one dimension from the other and the result will be the thickness of the shim which is required to be installed on the front face of the outer track of the input shaft bearing before the retainer and gasket are bolted into position.

75 Install the input shaft bearing retainer (complete with new oil seal - lips towards gearbox) making sure that the oil channel is at the top. Apply jointing compound to the gasket (both sides) before installing. Refit the clutch withdrawal components.

76 Reconnect the gearshift remote control, install the end cover brackets and the stay and refill the unit with the correct grade and quantity of oil.

6 Gearbox (5 speed type 235) - overhaul

1 Carry out the operations described in paragraphs 1 to 6, of the preceding Section.

2 Unscrew and remove the rear cover bolts.

3 Pressure must now be applied to the front of the input shaft so that as it is pressed to the rear, the end cover and housing will separate from the main gearbox casing. To provide this pressure, either tap the input shaft carefully with a soft-faced mallet or screw two lengths of studding into two of the front bearing retainer bolt holes and then using a piece of flat metal as a bridge piece, screw a bolt down onto the front end of the input shaft.

4 Repeat the operations described in paragraphs 18 to 22, of Section 5.

5 With the gears in the neutral position, remove the gearchange rod, plug, spring and locking plunger, also the speedometer pinion from the end cover.

6 Engage 5th gear, drive out the lock pin which secures the shift fork to the 4th/5th selector rod. Make sure that the pin can pass between the teeth of the gear below it.

7 Drive the 4th/5th selector rod out towards the front of the gearbox. Catch the detent balls which will be released.

8 Engage 3rd gear and drive out the lock pin which secures the shift fork to the 2nd/3rd selector rod. Make sure that the pin can pass between the teeth of the gear below it.

9 Drive the 2nd/3rd selector rod out towards the front of the gearbox. Catch the detent balls which will be released.

10 Move the selectors to neutral and then drive the hollow pins (located between the end cover and rear housing) out of the flange joint. Unscrew and remove the two tie bolts.

11 Pull the gear assembly together with the rear housing from the end cover. Catch the ejected detent balls.

5.52c Detent springs

5.54 Mainshaft and layshaft installed to end cover

5.55 Installing mainshaft rear bearing

5.59a Installing mainshaft rear bearing shim

5.59b Installing mainshaft rear bearing retainer

5.60 Installing coupling flange lockplate and nut to rear end of mainshaft

Fig. 6.49. Measuring mainshaft rear bearing track recess (A) on Type 232 gearbox

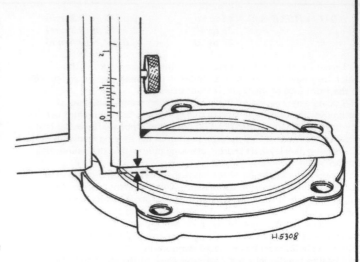

Fig. 6.50. Measuring height of mainshaft rear bearing retainer collar (Type 232)

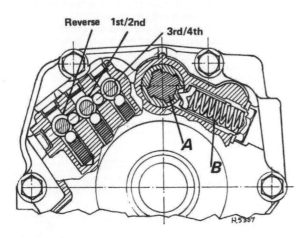

Fig. 6.51. Location of selector rod detent and interlock balls (Type 232 gearbox). Note engagement of gearchange rod taper bush (A) with plunger (B)

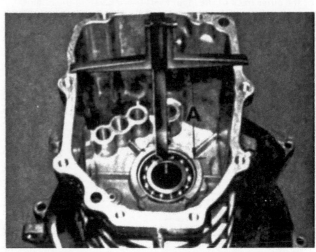

Fig. 6.52. Measuring depth (A) of input shaft bearing inner track from main gear casing flange (Type 232)

Fig. 6.53. Marking (B) on Type 232 input shaft and compensating shim (X)

5.64 Installing gearchange rod

5.65a Inserting a selector rod detent ball

5.65b Selector fork locking pins

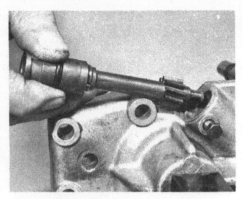

5.67 Installing speedo pinion

5.71a Joining end cover with geartrain to gearbox casing.

5.71b Installing shim in input shaft bearing recess.

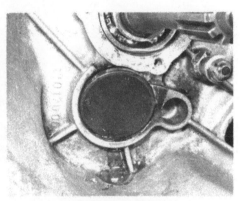

5.72 Input shaft bearing installed and layshaft bearing blanking plug being fitted

12 Secure the rear housing in a vice fitted with jaw protectors and from the rear end of the mainshaft, withdraw the speedometer drive gear and the needle bearing cage.

13 Drive the locking pins from the 1st/reverse selector rod dog and then withdraw the 1st/reverse selector rod from the rear of the housing and lift away the shift fork. At this stage, check the gearchange rod and dog for wear. If necessary, remove them after cutting the locking wire and unscrewing the lock bolt.

14 Extract 2nd/3rd and 4th/5th shift forks.

15 Using a soft-faced mallet, tap the rear end of the mainshaft towards the front of the gearbox and then using a two-legged puller, withdraw the 1st gearwheel, guide sleeve and spacer.

16 Carefully tap the mainshaft forward until 3rd gear synchro. unit butts against the 3rd gearwheel on the layshaft.

17 Unscrew the socket screws which retain the mainshaft roller bearing in the rear housing and extract the retainer plates.

18 The mainshaft and layshaft must now be pressed or tapped simultaneously from the rear housing while the housing is well supported on its front face.

19 Remove the input shaft from the front of the mainshaft and withdraw the synchro. sleeve and needle bearing cage.

20 Extract the circlip and withdraw the disc, shim, guide sleeve and 4th gearwheel with needle bearing cage.

21 Support the rear face of the 2nd gearwheel and press the mainshaft from the remaining components.

22 With the gearbox dismantled, check all components for wear or damage. Renew oil seals as a matter of routine. Inspect the synchro. units and renew any worn components, as described in Section 4, paragraph 32, but noting that the synchro. rings are identified by colour: 1st gear, green; 2nd/3rd gear, yellow; 4th/5th gear, white. The 1st gear ring is oval in shape.

23 Any wear in the mainshaft gearteeth will necessitate renewal of the mainshaft and layshaft gearwheels as a matched pair. Similarly any wear in the teeth of a layshaft gearwheel will necessitate the renewal of the matching mainshaft gearwheel.

24 If new components have been obtained then the mainshaft must be shimmed in the following way. Press 3rd gear with needle bearing, guide sleeve, 2nd gear with needle bearing, spacer bush and roller bearing onto the mainshaft. Now install the spacer bush of 1st gear (but without 1st gearwheel itself) onto the mainshaft. Slide on the guide sleeve and using a feeler blade, check the play between the spacer bush and the guide sleeve. Select a shim which equals this play in thickness and install it eventually on the rear face of the mainshaft bearing. When the 4th gear and other components have been fitted to the front end of the mainshaft, check the endfloat as shown and include a shim between guide sleeve and thrust washer to prevent it.

25 As previously explained, if the layshaft must also be fitted with some new components then a press will be required as the gearwheels are an interference fit. The gearwheels can be removed cold but when installing, heat them to between 248 and 302°F (120 to 150°C). Before 1st gear can be removed from the layshaft, remove the shaft end screw. When the new 1st speed gear has been installed to the layshaft, select a shim which will take up the gap between the thrust washer and the gear.

26 Commence reassembly by pressing the mainshaft and layshaft roller bearings into the rear housing. Secure the mainshaft bearing with the retaining plates.

27 Install the mainshaft and layshaft geartrains simultaneously into the rear housing bearings.

28 Install 1st gear to the rear end of the layshaft, fit the shim previously selected (paragraph 25), the thick washer and after applying Loctite to the threads of the securing screw, tighten it to the specified torque of 44 lb/ft (61 Nm). In order to prevent the layshaft turning as the screw is tightened, temporarily install the coupling flange to the rear end of the mainshaft and select 2nd gear. The coupling flange can now be held still with a suitable lever while the layshaft screw is tightened.

29 Install the previously selected shim (paragraph 24) onto the rear end of the mainshaft and fit the spacer bush.

30 Install 1st gear and guide sleeve noting that the longer ends of the guide bars are towards 1st gear.

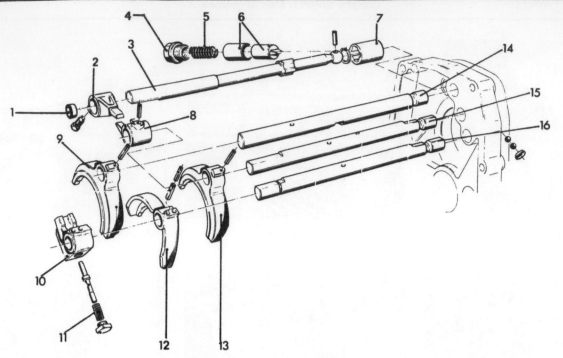

Fig. 6.54. Selector components (Type 235/5) gearbox

1 Lock bolt	5 Spring	9 4th/5th shift fork	13 2nd/3rd shift fork
2 Dog	6 Plunger	10 Dog	14 4th/5th selector rod
3 Gearchange rod	7 Taper bush	11 Plunger lock assembly	15 2nd/3rd selector rod
4 Threaded plug	8 Dog	12 1st/reverse shift fork	16 1st/reverse selector rod

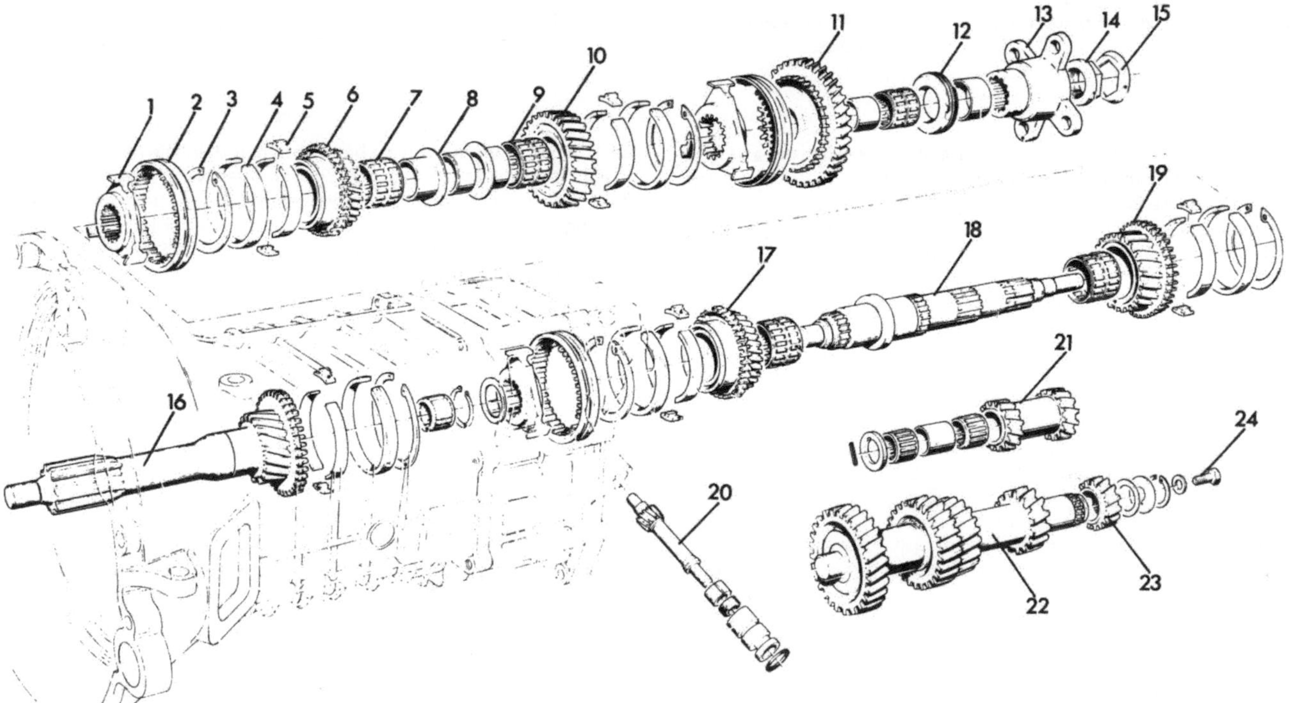

Fig. 6.55. Geartrain components (type 235/5 gearbox)

1 Guide sleeve	7 Needle roller bearing	13 Coupling	19 3rd gear
2 Synchro sleeve	8 Spacer sleeves	14 Coupling nut	20 Speedo drive gear
3 Circlip	9 Needle roller bearing	15 Locking plate	21 Double gear
4 Ring	10 1st gear	16 Input shaft	22 Layshaft
5 Key	11 Reverse gear	17 4th gear	23 1st gear (layshaft)
6 2nd gear	12 Speedo drive gear	18 Mainshaft	24 Layshaft screw

Fig. 6.56. Removing gearchange rod plunger from end cover
(Type 235/5)

Fig. 6.57. Removing speedo driven gear pinion from end cover
(Type 235/5)

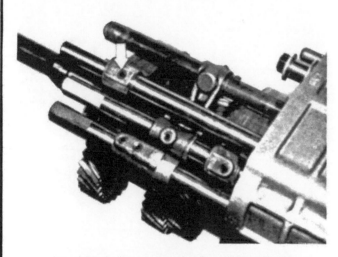

Fig. 6.58. 4th/5th shift fork lock pin (Type 235/5)

Fig. 6.59. Location of end cover to rear housing bolts and hollow pins
(Type 235/5)

Fig. 6.60. Removing speedo drive gear and needle cage (Type 235/5)

Fig. 6.61. 1st/reverse selector rod dog locking pins (Type 235/5)

Fig. 6.62. Withdrawing 1st/reverse selector rod, shift fork attached to synchro sleeve (Type 235/5)

Fig. 6.63. Withdrawing 1st gear from mainshaft (Type 235/5)

Fig. 6.64. Separating input shaft from front of mainshaft (Type 235/5)

Fig. 6.65. Withdrawing synchro guide sleeve and 4th gear from front of mainshaft (Type 235/5)

Fig. 6.66. Establishing mainshaft bearing shim thickness (Type 235/5)

Fig. 6.67. Establishing shim required to take up play at front end of mainshaft (Type 235/5)

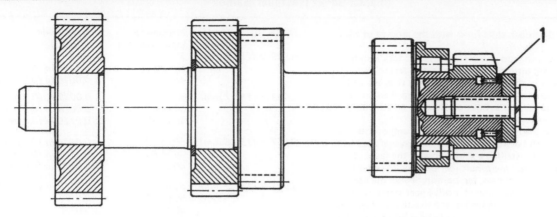

Fig. 6.68. Location of layshaft shim (Type 235/5) (1) shim

Fig. 6.69. Mainshaft and layshaft bearings in rear housing (Type 235/5)
Inset - 1 correct attitude of retaining plates, 2 incorrect

Fig. 6.70. Mainshaft and layshaft correctly installed into rear housing (Type 235/5)

Fig. 6.71. Fitting shim and spacer bush to rear end of mainshaft (Type 235/5)

Fig. 6.72. Correct installation of 1st gear and guide sleeve to rear end of mainshaft (Type 235/5)

31 Engage 2nd/3rd and 4th/5th shift forks with the grooves of their synchro. sleeves.
32 Install 1st/reverse selector rod complete with shift fork into the rear housing. Slide the dog onto the selector rod and secure with pins. Make sure that the seams in the pins are in alignment with the rod. If the gearchange rod or dog were removed, refit them in their correct relative position which is the rod cut-out downwards and the head of the locking bolt at the bottom.
33 Install reverse gear to the mainshaft. If any new components have been fitted, the tooth mesh between mainshaft and layshaft gears can be adjusted if necessary by installing a shim between the faces of the speedometer drive gear and the mainshaft rear bearing. In order to establish precisely the shim thickness, temporarily install the bearing to the rear cover and with the speedometer drive gear standing on the bearing, measure the distance of the face of the speedometer gear below the mating flange of the rear cover. This should be 0.866 ± 0.0039 in $(22.0 \pm 0.1 \text{ mm})$. Insert shims as necessary to adjust the speedometer gear face height. Remove the gear and bearing from the end cover.
34 From the end cover, unscrew the reversing lamp switch and prise out the sealing cap adjacent to it.
35 Install the double gear assembly and thrust washer onto the shaft in the end cover and using grease, stick a new gasket to the end cover mating flange.
36 Connect the end cover to the rear housing. Before the 1st/reverse selector rod is inserted into the end cover, the detent ball must be dropped into the hole vacated by the reversing lamp switch and kept depressed against its spring pressure using a small screwdriver.
37 Fit the two connecting bolts between the end cover and rear housing flanges.
38 Push the 1st/reverse selector rod to the neutral position.
39 Into the hole in the end cover from which the blanking cap was removed, insert an interlock ball and a detent ball. Hold the detent ball depressed and install the 2nd/3rd selector rod picking up the shift fork as it is installed. Fit a new fork locking pin (seam in alignment with selector rod).
40 Move 4th/5th synchro. sleeve to neutral. Insert an interlock ball and a detent ball into the 4th/5th selector rod hole in the end cover. Hold the detent ball depressed with a small screwdriver and install the 4th/5th selector rod, picking up its shift fork as it is inserted. Secure fork with a new lock pin (seam in alignment with selector rod).
41 Screw the reversing lamp switch into the end cover and tap in the sealing cap having applied jointing compound to it.
42 Install the hollow pins in the end cover to rear housing flanges.
43 Install the gearchange rod, plug, spring and pluinger also the speedometer pinion to the end cover.
44 Install the bush to the rear end of the mainshaft and then drive on the mainshaft rear bearing (identification numbers visible from the rear).
45 Determine the thickness of the rear bearing shim by following the procedure described in paragraph 47, 48 and 49, of Section 5. Install the shim, gasket and bearing retainer (complete with oil seal). Refit the coupling flange, lockplate and nut to the end of the mainshaft.
46 Measure the distance between the end face of the layshaft gear and the rear housing flange with a new gasket in position. Record this dimension (D).
47 Now measure the distance between the flange of the main gear casing and the face of the inner track of the layshaft bearing. Record this dimension (E).
48 To determine the thickness of the shim which must be placed on the end of the layshaft, subtract 'D' from 'E' and then subtract a further 0.0079 in (0.2 mm) to give the correct endfloat.
49 Measure the distance between the flange of the main gear casing and the face of the inner track of the input shaft bearing. Record this dimension (A).
50 Record the number (B) engraved on the input shaft and by using the following table, the thickness of the shim which must be installed on the input shaft can be established (see Fig. 6.53).

Depth (A)	Marking on input shaft (B)	Input shaft shim (C)
5.929 in (150.6 mm)	60 to 70	0
	50 to 60	0.0039 in (0.1 mm)
	40 to 50	0.0078 in (0.2 mm)
5.933 in (150.7 mm)	60 to 70	0.0039 in (0.1 mm)
	50 to 60	0.0078 in (0.2 mm)
	40 to 50	0.0118 in (0.3 mm)
5.937 in (150.8 mm)	60 to 70	0.0078 in (0.2 mm)
	50 to 60	0.0118 in (0.3 mm)
	40 to 50	0.0157 in (0.4 mm)
5.941 in (150.9 mm)	60 to 70	0.0118 in (0.3 mm)
	50 to 60	0.0157 in (0.4 mm)
	40 to 50	0.0196 in (0.5 mm)
5.945 in (151.0 mm)	60 to 70	0.0157 in (0.4 mm)
	50 to 60	0.0196 in (0.5 mm)
	40 to 50	0.0236 in (0.6 mm)
5.949 in (151.1 mm)	60 to 70	0.0196 in (0.5 mm)
	50 to 60	0.0236 in (0.6 mm)
	40 to 50	0.0276 in (0.7 mm)
5.953 in (151.2 mm)	60 to 70	0.0236 in (0.6 mm)
	50 to 60	0.0275 in (0.7 mm)
	40 to 50	0.0315 in (0.8 mm)
5.957 in (151.3 mm)	60 to 70	0.0275 in (0.7 mm)
	50 to 60	00315 in (0.8 mm)
	40 to 50	0.0315 in (0.8 mm)

51 Locate the main gear casing over the geartrain and secure it to the rear housing. Use a tubular drift to drive the inner track of the input shaft bearing onto the input shaft.
52 Repeat the operations described in paragraphs 62, 63, 64, 74, 75 and 76 of Section 5, of this Chapter.

7 Replacing 4 speed gearbox (type 232) with a 5 speed unit (type 235/5)

1 This can be carried out provided the following sequence of operations is followed.
2 Remove the propeller shaft (Chapter 7) and measure the distance between the centre of the gearbox output shaft and the underside of the transmission tunnel immediately above it.
3 Remove the existing gearbox and push out the top right-hand corner of the transmission tunnel to provide greater clearance for the new gearbox bracket. This re-shaping will have to be done by cutting and then welding in a new corner section of smaller radius.
4 Fit the modified clutch release components which are necessary with the new gearbox.
5 Cut the original rear mounting crossmember support brackets from the underside of the transmission tunnel.
6 Install the new gearbox and support its rear end on a jack so that the distance between the centre of the gearbox output shaft and the transmission tunnel immediately above it corresponds exactly with the measurement taken when the original gearbox was in position.
7 Fit the crossmember complete with flexible pads and new support brackets to the gearbox. Mark the position of the support bracket bolt holes and then remove the mounting assembly so that the bolt holes can be drilled. It will be found easier if pilot holes are drilled from below the car but the full-size holes drilled from inside the car.
8 Refit the crossmember and bolt the support brackets securely to the tunnel. Remove the jack.
9 Install the new modified propeller shaft which is essential with the new gearbox and change the speedometer driven components.
10 Fill the gearbox with oil.

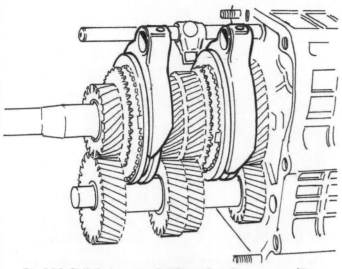

Fig. 6.73. Shift forks engaged with synchro sleeve grooves (Type 235/5)

Left-hand fork 3th/5th   *Right-hand fork 2nd/3rd*

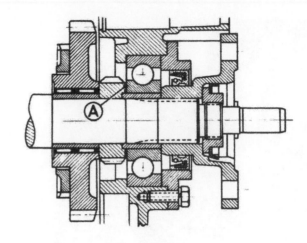

Fig. 6.74. Location of mainshaft shim to adjust mainshaft to layshaft gear tooth mesh (Type 235/5)

Fig. 6.75. Checking depth of speedo drive gear below end cover flange (Type 235/5)

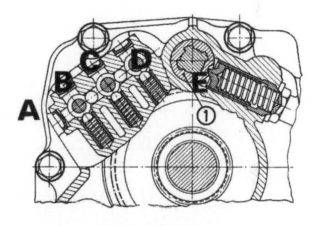

Fig. 6.76. Location of detent and interlock balls (Type 235/5)

A   1st/reverse   D   Gearchange rod taper bush
B   2nd/3rd   E   Plunger assembly
C   4th/5th

Fig. 6.77. Measuring length (D) of layshaft gear (Type 235/5)

Fig. 6.78. Measuring depth (E) of layshaft bearing below gear casing flange (Type 235/5)

Fig. 6.79. Location of layshaft adjusting shim (F) - Type 235/5

Fig. 6.80. Measuring depth (A) of input shaft bearing below gear casing flange (Type 235/5)

Fig. 6.81. Offering up new crossmember support brackets to underside of transmission tunnel after replacement of 4 speed gearbox with 5 speed type

8 Fault diagnosis - manual gearbox

Symptom	Reason/s
Ineffective synchromesh	Worn baulk rings or synchro hubs.
Jumps out of one or more gears (on drive or over-run)	Weak detent springs or worn selector forks or worn gears.
Noisy, rough, whining and vibration	Worn bearings and/or thrust washers (initially) resulting in extended wear generally due to play and backlash. Incorrectly shimmed components.
Noisy and difficult engagement of gears	Clutch fault (see Chapter 5).

Note: It is sometimes difficult to decide whether it is worthwhile removing and dismantling the gearbox for a fault which may be nothing more than a minor irritant. Gearboxes which howl, or where the synchromesh can be 'beaten' by a quick gearchange, may continue to perform for a long time in this state. A worn gearbox usually needs a complete rebuild to eliminate noise because the various gears, if re-aligned on new bearings will continue to howl when different wearing surfaces are presented to each other.

The decision to overhaul therefore, must be considered with regard to time and money available, relative to the degree of noise or malfunction that the driver has to suffer.

Chapter 6 Part 2:Automatic transmission

Contents

Specifications

Type ... ...   Zaknradfabrik Friedrickshafen 3HP-12 (028)

Ratios

1st	2.56 : 1
2nd	1.52 : 1
3rd	1.00 : 1
Reverse	2.00 : 1
Speedometer	2.50 : 1

Torque converter diameter ... ...   9.45 in. (240.0 mm) - number 204

Fluid capacity

Initial filling	8.2 pints - 4.65 litres - 4.9 US qts.
Oil changing	3.1 pints - 1.75 litres - 1.80 US qts.

Fluid type ... ...   Dexron

Shift speeds

	1st to 2nd	2nd to 3rd
Full throttle	23 to 27 mph (37 to 43 kph)	60 to 64 mph 97 to 103 kph)
'Kick-down'	37 to 41 mph (59 to 65 kph)	66 to 70 mph 106 to 112 kph)
	2nd to 1st	**3rd to 2nd**
	19 to 23 mph (31 to 37 kph)	39 to 43 mph (62 to 68 kph)

Torque wrench settings

	lb/ft	Nm
Starter switch locknut	25	35
Fluid drain plug	25	35
Torque converter to engine bolts:		
Small	18	25
Large	34	47
Output shaft/coupling flange nut	108	149
Propeller shaft front coupling nuts to output shaft flange	34	47

9 General description

1 This type of transmission may be optionally specified on certain models. It is fully automatic and incorporates a fluid filled torque converter and a planetary gear unit.

2 Three forward and one reverse speed are provided and a 'kick-down' facility for rapid acceleration during overtaking when an immediate change to a lower speed range is required.

3 Due to the complexity of the automatic transmission unit, if performance is not up to standard, or overhaul is necessary, it is imperative that this be left to a main agent who will have the special equipment and knowledge for fault diagnosis and rectification.

The contents of the following Sections are therefore confined to supplying general information and any service information and instruction that can be used by the owner.

10 Routine maintenance

1 The most important maintenance operation is the checking of the fluid level. To do this, run the car for a minimum of 5 miles (8 km),

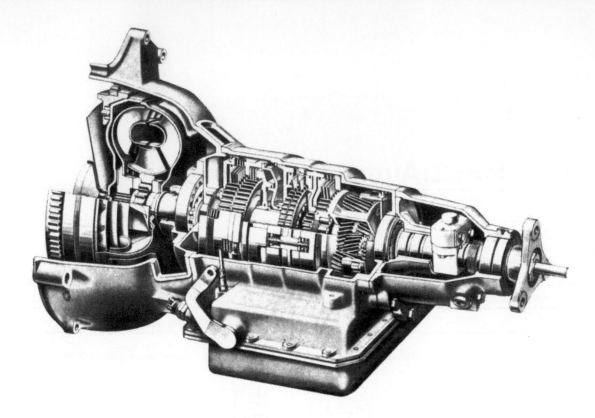

Fig. 6.82. 3HP-12 type automatic transmission

apply the handbrake and with the engine idling, move the speed selector lever to all positions, finally setting it in the 'P' detent.

2   With the engine still idling, withdraw the dipstick, wipe clean, re-insert it and withdraw it again for the second time. The fluid level should be between the 'low' and 'high' marks otherwise top it up to the correct level by pouring specified fluid down the combined filler/dipstick guide tube.

3   Occasionally, check the security of all bolts on the transmission unit and keep the exterior clean and free from mud or oil to prevent overheating.

4   Every 22000 miles (36000 km) drain the transmission fluid (hot) and refill with fresh fluid.

11 Automatic transmission - removal and installation

1   This operation is within the scope of the home mechanic where a new or reconditioned unit is to be installed or where a faulty unit must be removed for repair. It is emphasised however that the transmission should not be removed from the vehicle before the fault has been diagnosed under operational conditions by the repairing agent using special testing equipment.

2   Remove the lead from the battery negative terminal.

3   Disconnect the downshift cable and the throttle linkage.

4   Disconnect the oil filler tube support bracket and prise the tube from the transmission unit. Note the sealing 'O' ring and seal the opening with a piece of adhesive tape to prevent the entry of dirt.

5   Drain the transmission fluid by unscrewing the sump plug. If an oil cooler is fitted, disconnect the pipes.

6   Unscrew and remove all the bolts which secure the torque converter housing to the engine.

7   Detach the exhaust pipe support bracket from the rear end of the transmission.

8   Disconnect the exhaust downpipe from the manifold and then turn the steering to full left lock.

9   Disconnect the propeller shaft from the coupling flange at the rear of the transmission unit.

10 Disconnect the propeller shaft centre bearing and tie the propeller shaft to one side.

11 Disconnect the speedometer drive cable and the speed selector linkage from the transmission (see Fig. 1.7).

12 Identify and detach the leads from the starter inhibitor switch.

13 Loosen the reinforcement bracket bolts and remove the lower cover plate from the front of the torque converter housing (see Fig. 1.13).

14 Disconnect the driveplate from the torque converter, as described in Chapter 1, Section 8, paragraph 2 (Fig. 1.14).

15 Support the transmission on a jack, preferably of trolley type and then unbolt the rear mounting crossmember from the transmission and bodyframe.

16 Lower the jack carefully and withdraw the transmission to the rear. Use a piece of wood as a lever to keep the torque converter pressed into the converter housing during the withdrawal operation and expect some loss of fluid.

17 Installation is a reversal of removal but refer to Section 49, of Chapter 1, for correct engagement of the torque converter.

18 Refill the unit with the correct quantity and type of fluid and check the adjustment of the speed selector and accelerator linkage (see Sections 12 and 13).

19 Preload the propeller shaft centre bearing (see Chapter 7).

12 Speed selector linkage - adjustment

1   Disconnect the selector rod from the selector lever which is located on the left-hand side of the transmission unit.

2   Push the selector lever fully towards the front of the car and then move it rearwards three 'clicks' or detents.

3   Position the hand control lever in the neutral position and then offer the clevis fork on the front end of the selector rod to the selector lever on the side of the transmission casing. The clevis pin holes should be in perfect alignment. If not, release the locknut at the clevis fork at the rear of the rod and turn the rod as necessary.

4   When correctly installed there should be a free-movement at the hand control lever of 0.039 in (1.0 mm) when it is in the neutral mode.

5   Tighten the control rod locknut and check the speed selector in all positions at the same time turning the starter switch which should only operate in 'Neutral' or 'Park' otherwise the speed selector linkage or the inhibitor switch are incorrectly adjusted (see Section 14).

Fig. 6.83. Dipstick - automatic transmission

Fig. 6.85. Torque converter housing to engine bolts

Fig. 6.87. Starter inhibitor switch

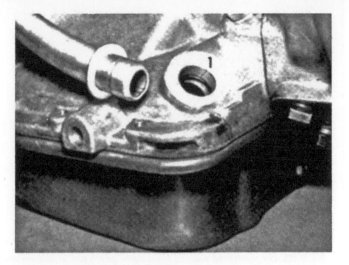

Fig. 6.84. Fluid filler/dipstick tube and 'O' ring seal (1)

Fig. 6.86. Exhaust pipe attachment to rear of transmission casing

13 Accelerator linkage and downshift cable - adjustment

1   Disconnect the downshift cable from the crank arm at the carburettor. Remove the air cleaner.

2   Have an assistant depress the accelerator pedal fully, compressing the 'kick-down' stop.

3   Check that the throttle valve plate is in the fully open (vertical) position by looking down the carburettor throat. If it is not, bend the stop to correct the attitude of the plate or adjust the length of the accelerator linkage (clevis fork and locknut) as appropriate.

4   With the assistant holding the accelerator pedal fully depressed, pull the downshift cable to its fully extended position. The hole in the clevis fork at the end of the cable should be in exact alignment with the hole in the crankarm on the throttle shaft. If not, adjust the cable clevis fork locknuts as necessary.

5   Reconnect the cable to the crankarm and test kickdown operation on the road.

14 Starter inhibitor switch - removal, refitting and adjustment

1   The starter inhibitor switch is also combined with the reversing lamp switch.

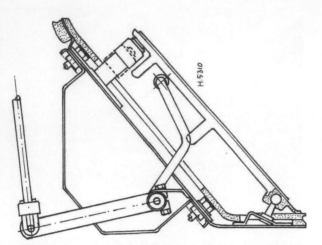

Fig. 6.88. Diagrammatic view of accelerator pedal and kickdown stop and linkage

Fig. 6.89. Downshift cable attachment to throttle rod

Fig. 6.90. Downshift cable locknuts

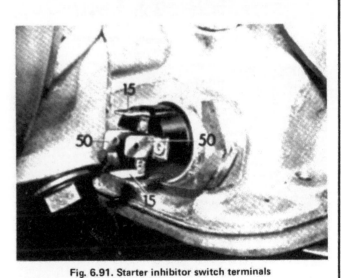

Fig. 6.91. Starter inhibitor switch terminals

Green/white and blue/black cables to 15.
Black/brown and brown to 50

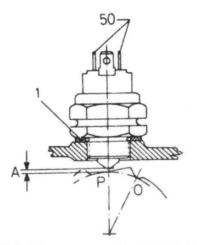

Fig. 6.92. Sectional view of starter inhibitor switch

1   Sealing washer   A   Switch plunger
                                          clearance

Fig. 6.93. Transmission rear coupling flange and flexible buffer (1)

2 Disconnect the leads having first identified them and the terminals to which they are connected.
3 Unscrew and remove the switch and extract the sealing washer.
4 Refitting is a reversal of removal but any failure of the switch to operate correctly may be due to an over-compressed sealing washer. A clearance 'A' between the end of the switch plunger and the actuating pawl must be maintained and this can only be precisely established by using a depth gauge to measure the distance of the pawl below the switch flange and then comparing this with the projection of the switch plunger. The thickness of the sealing washer can then be calculated to provide correct operation.

15 Rear oil seal - renewal

1 Provided the transmission fluid level is correct, any leaks from the rear of the transmission extension housing must be due to a faulty oil seal. This can be renewed without removing the transmission.
2 Disconnect the exhaust pipe from the rear of the transmission.
3 Disconnect the front of the propeller shaft from the transmission coupling flange.
4 Disconnect the propeller shaft centre bearing support from the bodyframe. Tie the propeller shaft to one side.
5 From the coupling flange, pull the flexible buffer and prise out the lockplate.
6 Holding the coupling flange quite still, unscrew and remove the nut and then pull off the flange.
7 The rear oil seal is now exposed and it may be extracted by levering or using a two-legged puller, the pressure from the centre screw being applied to the end of the transmission output shaft.
8 Drive in the new seal using a piece of tubing as a drift. The outer

face of the seal must lie 0.118 in (3.0 mm) below the end face of the extension housing.
9 Reassembly is a reversal of dismantling but preload the propeller shaft centre bearing, as described in Chapter 7.

16 Front oil seal - renewal

1 If loss of transmission fluid is evident at the base of the torque converter housing at the joint of the cover plate, then this is almost certainly due to a faulty front oil seal.
2 Access to this seal can only be gained if the engine or transmission is first withdrawn and the torque converter removed.
3 The oil seal may then be prised out with a lever or using a two-legged extractor applying pressure from its centre screw to the end of the transmission input shaft.
4 Drive in the new seal using a piece of tubing.
5 Installation is a reversal of removal and reassembly but make sure that the lugs of the torque converter are fully engaged with the primary oil pump (see Section 49, Chapter 1).

17 Speedometer driven gear pinion 'O' ring - renewal

1 Fluid leakage from this component can be rectified by unscrewing the lockbolt and withdrawing the pinion/gear assembly. Renew the 'O' ring seal.
2 If the internal seal is leaking (evident by fluid collecting in the lower loop of the speedometer cable) the pinion/gear assembly should be renewed.

Fig. 6.94. Speedometer driven gear/pinion 'O' ring (1) and lock bolt

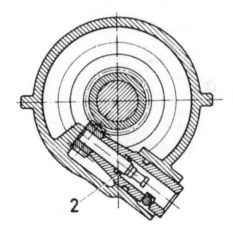

Fig. 6.95. Sectional view of speedometer gear/pinion showing internal oil seal

18 Fault diagnosis - automatic transmission

Symptom	Reason/s
Speed shifts too high or too low	Downshift cable incorrectly adjusted
No 'kick-down'	Downshift cable incorrectly adjusted
No forward or reverse drive	Low fluid level
Transmission slip	Low fluid level
	Downshift cable disconnected

Note: The faults listed are only those which are considered to be capable of rectification by the home mechanic.

Chapter 7 Propeller shaft

Contents

Specifications

| Type ... ... | Tubular, two section with centre bearing. Non-renewable universal joints or flexible couplings. Version with sliding sleeve. |

Torque wrench settings

	lb/ft	Nm
Shaft without sliding section		
Rear flange bolts	48	66
Flexible joint to transmission output flange	48	66
Shaft centre nut	130	179
Centre bearing support bolts	18	25
Shaft with sliding section		
Rear flange bolts	26	36
Flexible joint to transmission output flange	35	48
Threaded ring (sliding section)	29	40
Centre bearing support bolts	18	25
Shaft with flexible couplings throughout		
Coupling bolts	34	47
Threaded ring (sliding section)	29	40
Centre bearing support bolts	18	25

1 General description

1 The type of propeller shaft varies according to model and date of production. All shaft varies according to model and date of production. All shafts are of two section design with a centre bearing but some incorporate a sliding section while others do not.

2 All shafts are connected to the transmission output shaft flange by means of a flexible coupling.

3 Some shafts have centre and rear flexible couplings instead of conventional universal joints.

4 Renewal of the propeller shaft centre bearing and flexible couplings is possible but any wear in the universal joints will necessitate renewal of the complete shaft as the bearings are staked in position and the complete shaft is balanced to fine limits when manufactured. Always renew a propeller shaft with one of similar type to that fitted as original equipment. When dismantling, always mark the components so that they can be refitted in the same relative position.

2 Propeller shaft - removal and refitting

1 Disconnect the front silencer from the exhaust downpipe.

2 Unscrew and remove the bolts which secure the flexible coupling to the transmission unit output flange. Support the front end of the propeller shaft (photo).

3 Unscrew and remove the bolts which secure the propeller shaft rear flange to the pinion drive flange. Support the rear end of the propeller shaft.

4 Unscrew and remove the two bolts from the propeller shaft centre bearing. Lower and remove the shaft complete (photo).

5 Installation is a reversal of removal but observe the following points:

    a) *Hold the bolt heads on the flexible coupling quite still while the nuts are tightened so that the coupling does not become distorted.*

    b) *Use new locknuts on the rear flange bolts.*

    c) *Preload the centre bearing by pushing the support arms towards the front of the car by 0.08 in. (2.0 mm) before tightening the securing bolts.*

    d) *On propeller shafts which have a sliding section, the threaded ring should be left loose until the shaft has been installed and then tightened fully.*

6 A straight-edge and small blocks should be used to check the alignment of the two sections of the propeller shaft. Any misalignment can cause vibration or drumming during operation of the car and it can be corrected by inserting shims under the bearing support arms.

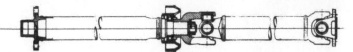

Fig. 7.1. Propeller shaft without sliding section

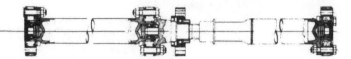

Fig. 7.2. Propeller shaft with sliding section

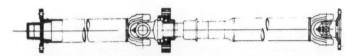

Fig. 7.3. Propeller shaft with sliding section and all flexible couplings

Fig. 7.4. Threaded ring on sliding type propeller shaft

2.2 Propeller shaft front coupling

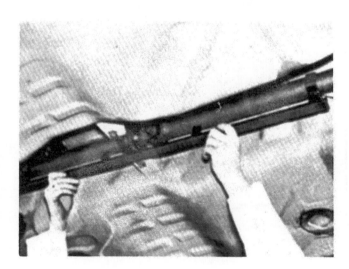

Fig. 7.5. Checking propeller shaft alignment

2.4 Propeller shaft centre bearing (non-sliding) type shaft

3 Flexible coupling - removal, overhaul and refitting

1 Removal of a flexible coupling is carried out more easily if the propeller shaft is first removed, as described in the preceding Section.
2 Extract the bolts which secure the coupling to the propeller shaft flange.
3 If the centring ring and ball socket are to be renewed, prise the sealing cap from the centre of the flexible coupling and extract the circlip.
4 Extract the internal components with a suitable tool.
5 When reassembling, pack the centring ring with high melting point Molybdenum grease.
6 A new flexible coupling is supplied complete with a band placed round its outer periphery. This is to slightly compress the rubber to enable the bolts to be inserted easily. When the coupling has been installed, cut through the band and discard it. If the original coupling is being refitted, a large worm-drive clip can be used as a substitute for the band.

Fig. 7.6. Extracting circlip from flexible coupling

Fig. 7.7. Flexible coupling centering ring components

1	Spring	4	Ball socket
2	Washer	5	Circlip
3	Centering ring	6	Sealing cap

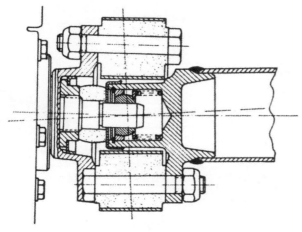

Fig. 7.8. Sectional view front flexible coupling

Fig. 7.9. Unscrewing propeller shaft centre nut

Fig. 7.10. Extracting centre bearing

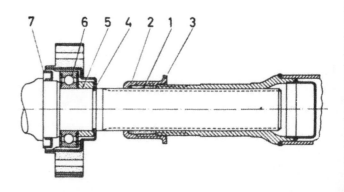

Fig. 7.11. Sectional view of centre bearing and sliding section

1	Felt ring	5	Collar
2	Washer	6	Ball bearing
3	Threaded ring	7	Dust deflector
4	Circlip		

4 Centre bearing - removal and installation (shafts without sliding section)

1   Remove the propeller shaft complete, as described in Section 2.
2   Mark the relative position of the two sections of the shaft and unscrew and remove the securing nut.
3   Pull off the centre bearing (leaving the dust deflector in position) using a two-legged puller.
4   Installation of the new bearing is a reversal of removal but smear the bearing seat with water or hydraulic fluid to faciliate fitting. Refit the propeller shaft by referring to Section 2.

5 Centre bearing - removal and installation (shafts with sliding section)

1   Disconnect the front silencer from the exhaust downpipe.
2   Disconnect the propeller shaft front flexible coupling from the transmission output flange.
3   Release the threaded ring on the sliding section and unbolt the centre bearing support.
4   Mark the relationship of the front section of the propeller shaft to the rear section and then withdraw the front section from the sliding sleeve.
5   Withdraw the centre bearing housing and then extract the bearing with a two legged puller.
6   Installation is a reversal of removal but remember to preload the centre bearing, as described in Section 2.

6 Fault diagnosis - propeller shaft

Symptom	Reason/s
Vibration	Incorrect alignment
	Worn coupling centre bearing
	Worn shaft centre bearing
	Worn or tight universal joints
	Flexible coupling deteriorated
Knock or 'clunk' on taking up drive or on overrun	Loose flange bolts
	Worn splines (sliding section)
	Worn universal joints

Chapter 8 Final drive unit and driveshafts

Contents

Specifications

Type	Hypoid bevel transmitting power through open driveshafts to independently sprung roadwheels.

Ratios

1500/1600	4.375 : 1 teeth 35/8
1502/1602	4.11 : 1 teeth 37/9 or 41/10
2000/2002	3.64 : 1 teeth 40/11
2002 TII	3.45 : 1 teeth 38/11

Oil capacity	1.6 Imp. pints; 0.9 litres; 0.95 US quarts

Torque wrench settings

	lb/ft	Nm
Final drive housing cover bolts	31	43
Pinion nut (minimum)	108	149
Bearing side cover bolts	18	25
Side driving flange centre bolts	80	111
Final drive housing to subframe bolts	56	77
Final drive to rear bracket bolts	38	53
Rear bracket to body bolts	38	53

1 General description

1 The final drive unit is of hypoid bevel gear type and it is mounted at its front end on the rear suspension subframe and at its rear end to a bodyframe support bracket.

2 The pinion housing may be of long or short type depending upon the date of production and model. Limited slip differential is available as an option.

3 Drive is transmitted to the roadwheels through open driveshafts which have either a needle roller type universal joint at the differential end and a needle roller sliding joint at the roadwheel end or a constant velocity (CV) joint at both ends. The latter requires no maintenance.

4 Due to the need for special tools and gauges it is recommended that only the operations described in this Chapter are carried out and where complete overhaul of the final drive unit is required that either this work is left to a BMW dealer or a new or reconditioned unit is used as a replacement.

2 Routine maintenance

1 Regularly inspect the condition of the flexible bellows which cover the driveshaft joints. If they are split, the driveshaft must be removed and new bellows installed, as described in Sections 8 or 9.

2 The oil in a new final drive unit should be changed after the first

1000 miles (1600 km) when it is warm after a run. Thereafter only topping-up is required at the specified servicing intervals.

3 Where sliding type driveshaft outer joints are used, every 20,000 miles (32,000 km) turn the rear roadwheels until the combined filler/drain plis pointing directly downwards. Unscrew the plug and drain the oil.

4 Turn the roadwheel until the hole is pointing upwards and inject 6.3 fluid oz. (180 cc) of EP 90 oil. Refit the plug.

5 Where needle roller type universal joints are fitted at the differential end of the driveshafts, apply the grease gun to their nipples every 5000 miles (8000 km).

3 Final drive unit - removal and installation

1 Refer to Chapter 7 and disconnect the propeller shaft from the pinion driving flange.

2 Mark the edges of the driveshaft inner flanges so that they can be reconnected in the same relative position and then unscrew the flange socket screws (photo).

3 Tie each of the driveshafts up out of the way.

4 Support the final drive unit on a jack (preferably of trolley type) and then unbolt the unit from the suspension subframe carrier.

5 Unbolt the rear support bracket from the bodyframe, then release the nuts which secure the final drive unit to the bracket and remove the bracket.

6 Lower the jack and withdraw the final drive unit from under the car.

7 Installation must be carried out in the following sequence to avoid any tendency to drumming or vibration when the car is operating.

a) Locate the final drive on the suspension subframe carrier and insert the securing bolts finger-tight.

b) Connect the propeller shaft.

c) Connect the rear support bracket to the final drive housing but leave the nuts finger-tight.

d) Bolt the rear support bracket to the bodyframe.

e) Allow the final drive unit to take up its own alignment within the limits of the four carrier bolt holes and then tighten these bolts and the two rear bracket bolts.

8 Reconnect the driveshafts, making sure that the marks made before disconnecting are in alignment.

9 The rubber bushes in the rear support bracket are not supplied as separate components and if the bushes become worn, the bracket must be renewed complete.

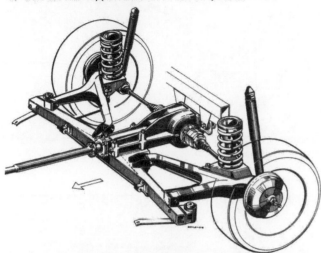

Fig. 8.1. Final drive (long pinion housing) and suspension on 2002

Fig. 8.2. Final drive (short pinion housing) and suspension on 2002 TII

Fig. 8.3. Driveshaft sliding type outer joint oil filler/drain plug

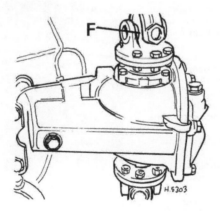

Fig. 8.4. Driveshaft inner joint grease nipple (F)

3.2 Driveshaft inner (CV) joint

Fig. 8.5. Final drive unit front mounting bolts

Fig. 8.6. Final drive unit rear mounting (3) support bracket

4 Driveshaft flange oil seals - renewal

1 Disconnect the driveshaft inner flange, as described in the preceding Section.
2 Hold the driving flanges of the final drive unit quite still, (this can be achieved by bolting a length of flat steel to two of the flanges bolt holes) and then unscrew the central securing bolt (photo).
3 Using a suitable two or three-legged puller, withdraw the driving flange from the side of the differential (photo).
4 The bearing cover oil seal can now be prised from its seat and the new one tapped into position so that the front face of the seal is 0.1575 in (4.0 mm) below the rim of its retainer.
5 Apply thread locking compound to the splines of the driving flange and refit it, tightening the central bolt to specified torque.
6 Reconnect the driveshaft.

4.2 Final drive flange and securing bolt

5 Pinion oil seal (short housing) - renewal

1 Remove the final drive unit from the car, as described in Section 3 and drain the oil.
2 Unbolt and remove the rear cover plate.
3 Hold each of the driveshaft driving flanges quite still, unscrew the central bolts and remove the flanges with a two or three-legged puller, as described in the preceding Section.
4 Mark each of the driving flange bearing cover/seal retainers in respect of which side of the final drive it is fitted and then unbolt it. Take care to retain the shim located under each bearing cover with its respective cover. Take the opportunity to renew the bearing cover 'O' rings.
5 Tilt the differential assembly to one side so that the taper roller bearing projects as far as possible from the bearing aperture and then lift the crownwheel and differential up and out of the final drive housing.
6 Mark the position of the driving flange to the pinion shaft.
7 Using a length of cord wound round the pinion diving flange and a spring balance attached to the end of the cord, determine the turning torque required to start the pinion rotating. Record this.
8 Hold the driving flange quite still (using a length of flat steel bolted to two of the flange holes) relieve the lockplate and unscrew the pinion nut.
9 Apply pressure to the end of the pinion shaft and press the shaft from the driving flange and roller bearings.
10 Prise out the oil seal and press in a new one, making sure that it is flush with the final drive housing. Fill the seal lips with grease.
11 Fit a new collapsible spacer and reassemble the pinion to the bearings and driving flange making sure that the latter is aligned correctly by mating the marks made before dismantling.
12 Hold the driving flange quite still, tighten the pinion nut to 108 lb/ft (149.0 Nm) and then using the cord and spring balance check the turning torque of the pinion. This should be the figure recorded before dismantling plus 25% to offset the drag of the new oil seal. Where the correct preload is not reached, tighten the pinion nut to a factionally higher torque wrench setting and recheck. Remember that if the preload of the pinion bearings is set too high, it cannot be reduced by backing off the pinion nut. In this case, the collapsible spacer will have to be renewed for the second time and the adjustment procedure carried out all over again.

4.3 Removing a final drive flange

6 Pinion oil seal (long housing) - renewal

1 The pinion locknut need not be disturbed when renewing the pinion oil seal on this design of the final drive unit and the unit need not be removed from the car and dismantling is therefore reduced to the following operations.
2 Disconnect the rear end of the propeller shaft.
3 Unscrew the pinion driving flange nut and draw off the flange using a small to or three-legged puller.
4 Prise the oil seal from its location and drive in the new one so that it is recessed by 0.24 in. (6.0 mm) as shown. Apply grease to the seal lips and reassemble in the reverse order to dismantling.

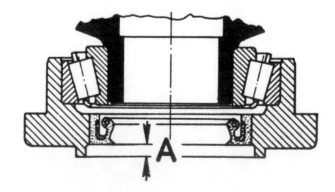

Fig. 8.7. Bearing side cover oil seal installation diagram

A = 0.1575 in (4.0 mm)

Fig. 8.8. Applying locking compound to splines of the driving flange

Fig. 8.9. Bearing side cover ('O' ring arrowed)

Fig. 8.10. Withdrawing differential assembly from housing

Fig. 8.11. Pinion to driving flange alignment marks

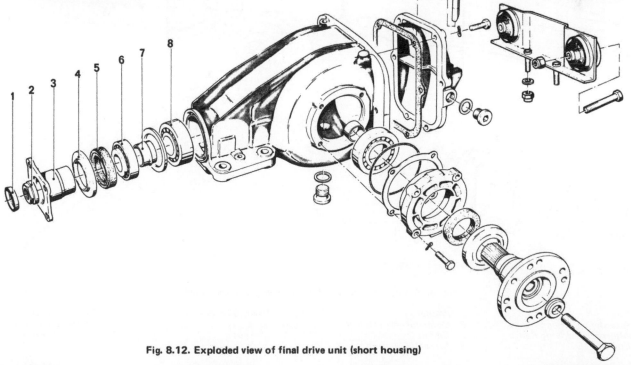

Fig. 8.12. Exploded view of final drive unit (short housing)

1 Lockplate	3 Driving flange	5 Oil seal	7 Collapsible sleeve
2 Pinion nut	4 Dust deflector	6 Taper roller bearing	8 Taper roller bearing

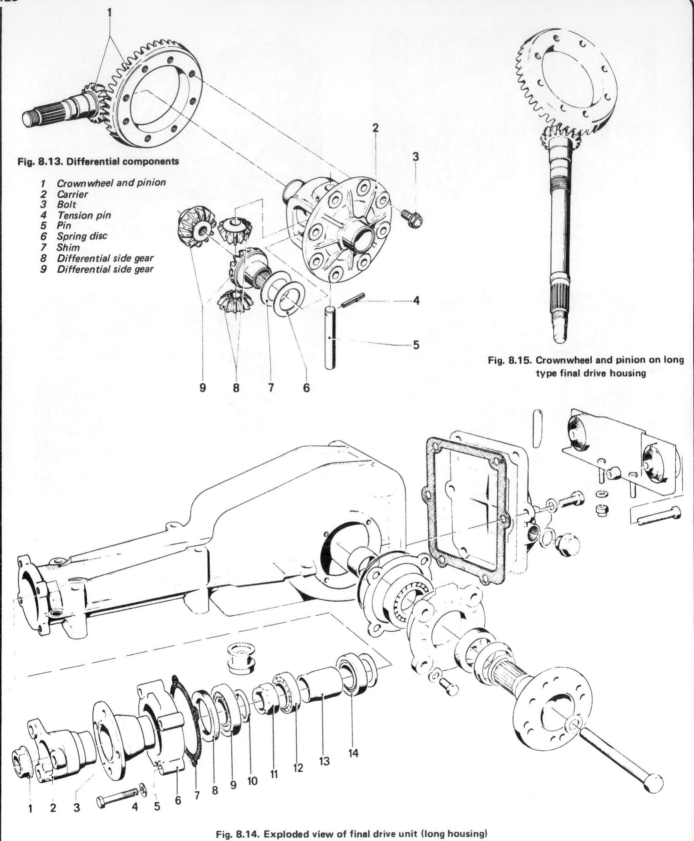

Fig. 8.13. Differential components

1   Crownwheel and pinion
2   Carrier
3   Bolt
4   Tension pin
5   Pin
6   Spring disc
7   Shim
8   Differential side gear
9   Differential side gear

Fig. 8.15. Crownwheel and pinion on long
type final drive housing

Fig. 8.14. Exploded view of final drive unit (long housing)

1 Nut	5 Washer	9 Taper roller bearing	12 Taper roller bearing
2 Alternative driving flange	6 Cover/oil seal retainer	10 Collar	13 Collapsible spacer
3 Alternative driving flange	7 Gasket	11 Locknut	14 Taper roller bearing
4 Bolt	8 Oil seal		

3 Pull off the sliding joint housing and withdraw the mushroom shaped caps. Apply thick grease to the ends of the needle rollers to prevent them being displaced and refit the caps.
4 Withdraw the narrower end of the bellows over the bearing ring assemblies.
5 Refitting the new bellows is a reversal of removal but positin the clips in their original positions in order to maintain the balance of the driveshaft.

9 Flexible bellows (constant velocity type joint) - renewal

1 Prise out the sealing cover from the end of the joint (photo).
2 Extract the circlip (photo).
3 Remove the bands which secure the bellows (mark the position of their turned over ends first to maintain shaft balance) (photo).
4 Support the joint and press the driveshaft from it (photo).
5 Slide the bellows from the driveshaft (photo).
6 The joint can be dismantled by tiling the inner components and extracting the balls from their grooves (photo).
7 To refit, first pack the joint and bellows with grease and locate the bellows on the driveshaft.
8 Check that the concave side of the clamp ring faces the joint. Press the driveshaft into the joint.
9 Clean the contact surfaces of shaft and joint and bellows. Apply Bostik sealant and secure the bellows with clamp bands. These bands should be drilled so that they can be compressed with a pair of round-nosed pliers before bending over the ends of the bands.

10 Driveshaft inner joints - servicing

1 Remove the driveshaft and extract the circlips from the outer end of each of the needle bearing cups.
2 Mark the relationship of the flange to the driveshaft yoke so that they can be refitted in the same relative position.
3 Place a socket slightly smaller in diameter than a bearing cup on the end face of one of the cups and another socket large enough to receive a bearing cup on the face of the yoke opposite. Apply pressure in the

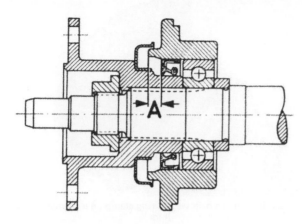

Fig. 8.16. Pinion oil seal (long housing) installation diagram

A = 0.24 in (6.0 mm)

7 Driveshafts - removal and installation

1 This is simply a matter of unscrewing the socket headed screws from the inner and outer flanges (first having marked the edges of the flanges for exact refitting) and lifting the shaft away.
2 During removal or installation, support the shaft so that the joints at either end are not strained beyond their maximum bending angles otherwise they may be damaged or disconnected.

8 Flexible bellows (sliding type joint) - renewal

1 Drain the oil, as described in Section 2, paragraph 3.
2 Mark the position of the bellows clamps and then remove them.

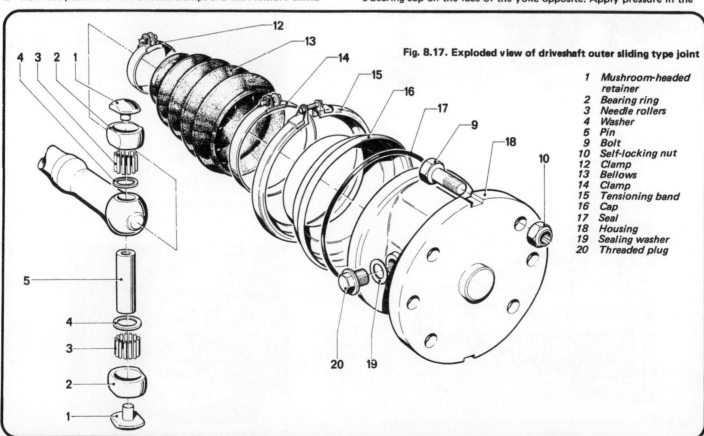

Fig. 8.17. Exploded view of driveshaft outer sliding type joint

1 Mushroom-headed retainer
2 Bearing ring
3 Needle rollers
4 Washer
5 Pin
9 Bolt
10 Self-locking nut
12 Clamp
13 Bellows
14 Clamp
15 Tensioning band
16 Cap
17 Seal
18 Housing
19 Sealing washer
20 Threaded plug

jaws of a vice and one cup will be partially ejected. Grip the ejected
cup with grips and twist it from the yoke and extract the needles. Turn
the joint through 180° and using the smaller socket to apply pressure
to the end of the spider from which the cup has already been removed,
eject the opposite pair of bearing cups.

4   To fit the new universal joint repair kit (spider, cups and needle
bearings) press in one cup only half way into the yoke of the driveshaft
and insert the needle bearings holding them in position with grease.
Insert the spider and locate the opposite cup complete with needles
again retained with grease. Press both bearing cups fully home in the
jaws of a vice making sure that the needles do not become trapped. Use
the narrow socket to depress each bearing cup far enough in the yoke to
expose the circlip groove.

5   Position the grease nipple or grease nipple hole towards the
driveshaft and repeat the bearing installation operations on the opposing
pair having aligned the drive flange yoke so that the marks made before
dismantling are correctly positioned.

11 Fault diagnosis - final drive unit and driveshafts

A noisy differential unit will necessitate its removal and overhaul
or its exchange for a reconditioned unit. Before taking this action, check
the following components as it is possible for sounds to travel and
mislead the owner as to the source of the trouble.

Check for:

Loose driveshaft flange
Tyres out of balance
Dry or worn rear hub bearings (see Chapter 11)
Worn differential flexible mountings
Lack of oil in outer sliding joints of driveshafts
Damaged or worn driveshaft universal joint
Worn or seized driveshaft ball grooves
Loose roadwheel nuts

Fig. 8.20. Installing a bellows securing clamp on a constant velocity
joint

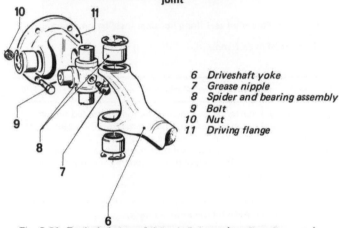

6   *Driveshaft yoke*
7   *Grease nipple*
8   *Spider and bearing assembly*
9   *Bolt*
10 *Nut*
11 *Driving flange*

Fig. 8.21. Exploded view of driveshaft inner (needle roller type)
universal joint

Fig. 8.18. Constant velocity joint components

1   *Sealing cover*   3   *Joint housing*
2   *Circlip*   4   *Bellows*

9.1 Removing cover from CV joint

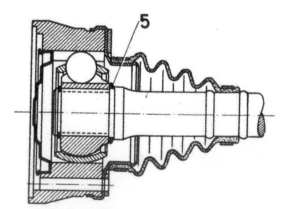

Fig. 8.19. Sectional view of constant velocity joint showing clamp
ring (5)

9.2 Extracting circlip from CV joint

9.3 Bellows securing bands on CV joint

9.4 Releasing CV joint from driveshaft

9.5 Removing bellows from driveshaft

9.6a Dismantling CV joint

9.6b CV joint outer ring

9.6c CV joint cage and centre track

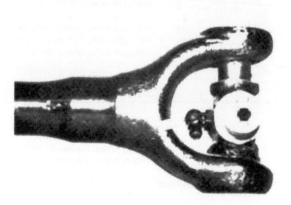
Fig. 8.22. Correct location of driveshaft universal joint nipple

Chapter 9 Braking system

Contents

Specifications

System type	Four wheel hydraulic with vacuum servo. Discs front, drums rear. Handbrake mechanical to rear wheels

Disc brakes

	1500/1600	1502/1602	2000 Touring	2002	2002 TI and TII
Diameter of disc	10.55 in (268.0 mm)	9.45 in (240.0 mm)	10.71 in (272.0 mm)	9.45 in (240.0 mm)	10.08 in (256.0 mm)
Minimum disc thickness	0.335 in (8.5 mm)	0.374 in (9.5 mm)	0.461 in (11.7 mm)	0.354 in (9.0 mm)	0.459 in (11.7 mm)
Maximum disc run-out	0.008 in (0.2 mm)	0.008 in (0.2 mm)	0.008 in (0.2 mm)	0.008 in (0.2 mm)	0.008 in (0.2 mm)
Minimum disc pad thickness (wear limit)	0.079 in (2.0 mm)	0.079 in (2.0 mm)	0.079 in (2.0 mm)	0.079 in (2.0 mm)	0.079 in (2.0 mm)

Drum brakes

	1500/1600	1502/1602	2000 Touring	2002	2002 TI and TII
Diameter of drum	9.84 in (250.0 mm)	7.87 in (200.0 mm)	9.84 in (250.0 mm)	9.06 in (230.0 mm)	9.06 in (230.0 mm)
Maximum out of round	0.004 in (0.1 mm)	0.004 in (0.1 mm)	0.004 in (0.1 mm)	0.004 in (0.1 mm)	0.004 in (0.1 mm)
Maximum internal oversize diameter ...	+ 0.059 in (1.5 mm)	+ 0.059 in (1.5 mm)	+ 0.059 in (1.5 mm)	+ 0.059 in (1.5 mm)	+ 0.059 in (1.5 mm)
Shoe width	1.575 in (40.0 mm)	1.575 in (40.0 mm)	1.575 in (40.0 mm)	1.575 in (40.0 mm)	1.575 in (40.0 mm)
Minimum lining thickness (wear limit)	0.12 in (3.0 mm)	0.12 in (3.0 mm)	0.12 in (3.0 mm)	0.12 in (3.0 mm)	0.12 in (3.0 mm)

Single circuit

	1500/1600	1602	2000 Touring	2002
Master cylinder, diameter of piston	11/16 in (17.46 mm)	3/4 in (19.05 mm)	15/16 in (23.81 mm)	0.8126 in (20.64 mm)
Caliper piston diameter	1.89 in (48.0 mm)	1.89 in (48.0 mm)	2.13 in (54.0 mm)	1.89 in (48.0 mm)
Rear wheel cylinder, diameter of pistons	5/8 in (15.87 mm)	11/16 in (17.46 mm)	11/16 in (17.46 mm)	11/16 in (17.46 mm)

Dual circuit

	1502/1602	2000 Touring	2002	2002 TI and TII
Tandem master cylinder diameter of pistons	0.812 in (20.64 mm)	0.937 in (23.81 mm)	0.812 in (20.64 mm)	0.9374 in (23.81 mm)
Caliper piston diameter:				
Two pistons	1.89 in (48.0 mm)	Not applicable	1.89 in (48.0 mm)	Not applicable
Four pistons	1.339 in (34.0 mm)	1.575 in (40.0 mm)	1.339 in (34.0 mm)	1.575 in (40.0 mm)
Rear wheel cylinder, diameter of pistons	11/16 in (17.46 mm)	11/16 in (17.46 mm)	5/8 in (15.87 mm)	11/16 in (17.46 mm)

Torque wrench settings

	lb/ft	Nm
Caliper securing bolts	70	97
Disc to hub bolts	48	66

1 General description

The braking system is of four wheel hydraulic type. Disc brakes are fitted to the front wheels and drum brakes to the rear. The handbrake operates through cables to the rear wheels only.

A number of different layouts and components may be encountered dependent upon date of vehicle production, model and operating territory. The hydraulic circuit may be of single or dual type. All systems have servo assistance but the vacuum booster itself may be connected directly to the foot pedal or single or twin boosters may be remotely sited within the engine compartment dependent upon the circuit layout employed.

The front disc brakes require no adjustment and the calipers on all models are of fixed type but they may be of two or four cylinder construction dependent upon the type of hydraulic circuit.

The rear drum brakes require regular adjustment and this adjustment automatically takes up any slack in the handbrake.

Various additional features are incorporated in the braking system according to model and operating territory and these include a fluid pressure warning device and a special spring in the front calipers which causes a need for higher pedal pressure when the disc pad friction material has worn to its minimum specified thickness.

The operations described in this Chapter apply to all types of components which may be encountered but the construction may differ in detail from those illustrated. Always renew a component with one of precisely similar type.

2 Rear brakes - adjustment

1 At the intervals specified in 'Routine Maintenance' jack-up each of the rear roadwheels.
2 Chock the front wheels and relase the handbrake.
3 On each brake backplate two hexagon-headed adjusters are located. When viewed from the direction of the differential, turn the left-hand one anticlockwise until the wheel is locked and then back the adjuster off until the wheel can be turned without binding.
4 Now turn the right-hand adjuster clockwise until the wheel is again locked and then back off the adjuster until the wheel can be turned without binding.
5 Lower the car and apply the handbrake. The adjustment of the rear brake shoes will normally also adjust the handbrake but where the travel of the handbrake lever is still excessive due to cable stretch, further adjustment can be carried out, as described in the next Section.

3 Handbrake - supplementary adjustment

1 Jack-up the rear roadwheels having first chocked the front roadwheels securely.

2 Push back the rubber gaiter from around the handbrake lever and then pull the lever on four notches of its ratchet.
3 Holding the threaded rod at the front end of one cable quite still with a pair of pliers, release the locknut on the rod and then turn the adjusting nut down the rod until the roadwheel is just locked.
4 Repeat the operation on the second cable.
5 Release the handbrake fully and check that both the rear roadwheels turn freely without binding.
6 Carry out any slight readjustment of the nuts to equalise the action and then retighten the locknuts, refit the gaiter and lower the car to the ground.

4 Front disc pads - inspection and renewal

1 Jack-up the front of the car and remove the roadwheels (photo).
2 Extract the two retaining pins and the anti-rattle spring (photo).
3 At this stage, inspect the thickness of the disc pad friction material. If it has worn to 0.08 in (2.0 mm) or less than the pads should be renewed on both front brakes as an axle set (photo).
4 Withdraw the pads by gripping their ends with a pair of pliers.
5 Brush any dust from the ends of the pistons and the pad recess (photo).
6 Using a syphoning device such as a poultry baster, withdraw some of the fluid from the brake fluid reservoir. This is done to accommodate additional fluid which will be displaced when the caliper pistons are depressed, as described in the next paragraph.
7 Using a flat piece of wood or metal, depress the caliper pistons into their cylinders in order to accept the new thicker disc pads. Keep the pistons square and depress them until they reach their stops. The reservoir fluid level will rise during this operation, make sure that it does not overflow.
8 Install the new pads, the spring and the pins and then apply the foot brake hard several times and finally top-up the fluid reservoir.
9 Refit the roadwheels and lower the car to the ground.

5 Rear brake shoes - inspection and renewal

1 Jack-up the rear of the car and remove the roadwheels.
2 Release the handbrake fully and slacken the shoe adjusters right off.
3 Remove the drum. If it is tight, tap it off using a block of hardwood between the hammer and the drum.
4 Brush away all dust from the shoes and the interior of the drum - **do not inhale this dust.**
5 Inspect the linings, if the rivets have worn down to or nearly down to the rivets, the shoes must be renewed. If bonded type linings are fitted, then the shoes must be renewed if the thickness of the friction material has worn to 0.08 in (2.0 mm) or less.
6 Prise the heavy shoe spring from the holes at the bottom end of the two brake shoes.

7   Pull both shoes outwards at their upper ends and unhook the shoe return spring.

8   Lower the shoes and disconnect the handbrake cable from the shoe lever. Extract the strut from between the two shoes.

9   Obtain new or reconditioned shoes which have factory fitted linings, it is not worth attempting to reline shoes yourself.

10 Arrange the new shoes in their correct position with regard to leading and trailing ends and transfer the handbrake lever to the new trailing shoe.

11 Engage the end of the handbrake cable with the bottom of the shoe lever and then offer both shoes to the wheel cylinder making sure that the strut which runs between the two shoes is so positioned that the longer side is nearest you and towards the shoe lever (photos).

12 The upper shoe return spring must be connected so that the longer end of the spring is attached to the shoe which carries the handbrake lever.

13 Install the heavy lower spring so that it is located behind the anchor plate.

14 Refit the drum, adjust the brake shoes, as described in Section 2, install the roadwheel.

15 Repeat all the foregoing operations on the opposite brake and then lower the car to the ground.

6   Caliper - removal and refitting

1   If the caliper is to be withdrawn only so that the hub or disc can be removed, do not disconnect the hydraulic lines but detach the fluid line bracket from the suspension strut, unbolt the caliper and move it to one side and tie it up out of the way.

If the caliper is to be removed for overhaul or renewal then in order to minimise loss of hydraulic fluid when the fluid lines are disconnected, remove the reservoir cap and stretch a piece of plastic sheeting over the filler neck and then screw the cap over the sheeting and screw the cap on again. This will create sufficient vacuum to prevent the fluid running out.

2   Jack-up the roadwheel and remove it.

3   Extract the disc pads.

4   Disconnect the fluid lines from the caliper by unscrewing the unions.

5   Unscrew and remove the two caliper securing bolts and withdraw the caliper from the disc.

6   Refitting is a reversal of removal but on completion, bleed the brake hydraulic stem, as described in Section 23. Remember to remove the plastic sheeting from the reservoir filler cap.

Fig. 9.1. Adjusting a rear brake

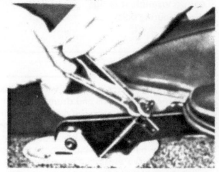

Fig. 9.2. Adjusting a handbrake cable

4.1 Front disc brakes

4.2 Disc pad retaining pins and anti-rattle spring

4.3 Removing a disc pad

4.5 Caliper showing dust excluders

5.11a Rear brake shoe assembly

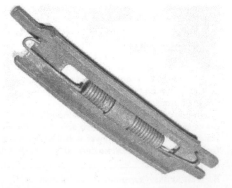

5.11b Rear brake shoe strut and return spring

7 Caliper (two piston type) - overhaul

1 With the caliper removed, carefully withdraw the dust excluders from the ends of the pistons.
2 Hold one of the pistons depressed and apply air from a tyre pump to the fluid inlet hole. This will eject the piston not being held.
3 Prise out the piston seal from the groove in the cylinder taking great care not to scratch the metal surfaces. Discard the seal and obtain a repair kit.
4 Fit the new seal using the fingers only to manipulate it, then dip the piston in clean hydraulic fluid and insert it squarely into the cylinder bore.
5 Hold this reconditioned piston depressed, apply air pressure to the fluid inlet hole and eject the second piston.
6 Repeat the operations to renew the seal and then insert the piston into its bore.
7 Install new dust excluders to the ends of the pistons.
8 If during the overhaul operations, any scratches, scoring or 'bright' wear areas are observed on the surfaces of the piston or cylinder bores, the caliper must be renewed complete.

8 Caliper (four piston type) - overhaul

1 With the caliper removed, the overhaul operations are similar to those described for two piston types in the preceding Section.
2 Work on two opposing pistons at a time and make sure that the pistons are not interchanged as they must be installed in their original bores. Use pieces of masking tape or a spirit marker to identify the pistons and cylinder, do not scratch or dot punch marks on them.
    The following points must be noted when overhauling a caliper unit. Always use hydraulic fluid or methylated spirit for cleaning - noting else. Never separate the two halves of the caliper by unscrewing the connecting bolts. Remember that hydraulic fluid is an effective paint stripper and do not spill any on the bodywork.

9 Front brake disc- examination, removal and refitting

1 Jack-up the front of the car and remove the roadwheel and disc pads.
2 Examine the surface of the disc for deep scoring or grooving. Light scoring is normal but anything more severe should be removed by taking the disc to be surface ground provided the thickness of the disc is not reduced below specification, otherwise a new disc will have to be fitted.
3 Check the disc for run-out. To do this, a dial guage will be needed although a reasonable check can be made using feeler blades between the face of the disc and a fixed point. Turn the disc slowly with the hand and if the run-out exceeds 0.008 in. (0.2 mm) then the disc must be renewed.
4 To remove the disc, first remove the hub assembly, as described in Chapter 11 and then unscrew the socket headed screws and lift the disc from the hub.
5 If necessary, the disc inner guard plate can be removed after unscrewing the three securing bolts.
6 Refitting is a reversal of removal.

10 Rear wheel cylinders - removal and refitting

1 Jack-up the rear of the car and remove the roadwheel and the brake drum.
2 Turn the shoe adjusters to their maximum adjustment position.
3 To minimise loss of hydraulic fluid, place a piece of plastic sheeting under the fluid reservoir cap and screw the cap on tightly.
4 Disconnect the fluid inlet pipe from the wheel cylinder and then unscrew the bleed screw and the cylinder securing bolts.
5 Prise the upper ends of the brake shoes slightly apart and withdraw the wheel cylinder (photo).
6 Refitting is a reversal of removal but adjust the shoes (Section 2) and after removing the plastic sheeting, bleed the hydraulic system, as described in Section 23.

11 Rear wheel cylinder - overhaul

1 Remove the cylinder, as described in the preceding Section, and brush and clean away all external dirt.
2 Remove the two rubber boots and extract the internal components. If the pistons are stuck, air pressure can be applied to the fluid inlet hole provided the bleed screw is refitted and closed.
3 Wash all components in hydraulic fluid or methylated spirit and examine the surfaces of the pistons and cylinder bores for scoring, scratching or 'bright' wear areas. If these are evident, renew the wheel cylinder complete.
4 If the pistons and cylinder are in good condition, discard the old seals and boots and obtain a repair kit.
5 Fit the new seals using the fingers only to manipulate them into position.
6 Dip the pistons in clean fluid before assembling.

12 Brake drum - inspection and renovation

1 With the brake drum removed, tap it with a spanner and note if there is a clear ring. If not the drum may be cracked.
2 Deep grooves or scored internal surfaces are caused by rivets due to non-renewal of worn linings. If the brake drum has worn oval in shape then the amount of out-of-round can only be accurately checked using an internal type vernier or dial gauge. The maximum out-of-round and the amount by which the internal diameter of the drum can be increased by re-finishing are to be found in the 'Specifications' Section.

13 Master cylinder (with remote servo unit) - removal and refitting

1 Syphon the fluid from the reservoir and disconnect the supply hose to the master cylinder (photo).
2 Disconnect the leads from the stop lamp switch if located on the master cylinder. On some models, the switch may be at the booster or attached to the brake pedal bracket.
3 Disconnect the fluid lines from the master cylinder by unscrewing the unions.
4 Remove the accelerator pedal (see Chapter 3) and peel back the carpet from around the brake pedal.
5 Detach the pedal return spring and the pushrod from the foot pedal arm.
6 Remove the securing bolts from the master cylinder flange and withdraw the cylinder into the engine compartment.
7 Refitting is a reverse of removal but bleed the brakes on completion (see Section 23).
8 Check that the brake pedal has a slight free-movement so that the master cylinder pushrod is not under tension when the pedal is fully released. Any adjustment needed can be carried out by slackening the clevis fork locknut and rotating the pushrod.

14 Master cylinder (with attached servo unit) - removal and refitting

1 Remove the air cleaner.
2 Syphon the fluid from the master cylinder reservoir and disconnect the supply hoses from the cylinder (except on North American versions which have integral reservoirs) (photo).
3 Disconnect the fluid pressure lines from the master cylinder. If there is any liklihood of confusing the pipe connections on refitting, mark them.
4 Unscrew and remove the nuts which secure the master cylinder to the front of the servo booster unit.
5 When refitting, check the condition of the 'O' ring seal between the flange of the master cylinder and the servo unit. Renew the seal if necessary. On tandem type master cylinders, a clearance must be maintained between the end of the pushrod and the master cylinder piston of 0.002 in (0.5 mm). Insert shims if necessary behind the pushrod domed nut.

Fig. 9.3. Caliper securing bolts and fluid lines

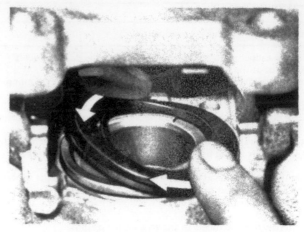

Fig. 9.4. Removing dust excluder from caliper piston (two cylinder type)

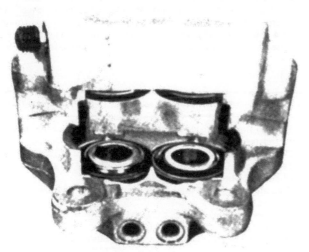

Fig. 9.5. Four cylinder type caliper

Fig. 9.6. Extracting a piston seal (four cylinder type caliper)

Fig. 9.7. Disc securing screws

Fig. 9.8. Disc guard plate and securing bolts (1)

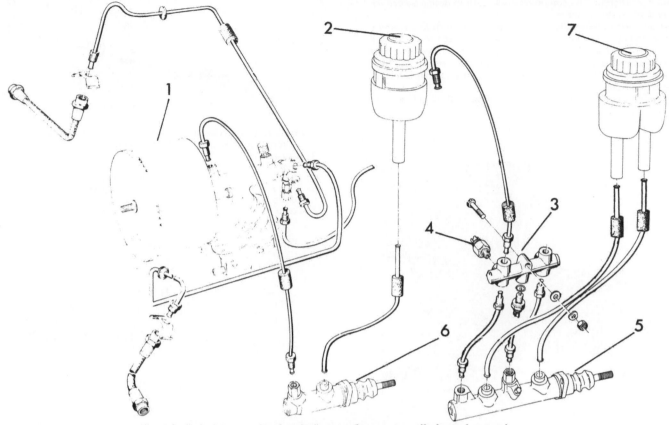

Fig. 9.9. Typical layout showing single or tandem master cylinder and remotely mounted servo unit

1 Servo unit
2 Master cylinder reservoir (single type)
3 Pressure differential switch
4 Stop lamp switch
5 Tandem type master cylinder
6 Single type master cylinder
7 Master cylinder fluid reservoir (tandem type)

10.5 Rear brake backplate and wheel cylinder

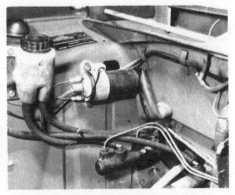

13.1 Tandem type waster cylinder and fluid reservoir

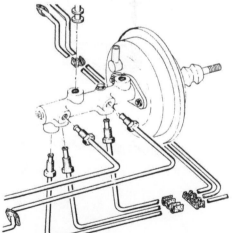

Fig. 9.10. Master cylinder with attached servo but remote reservoir

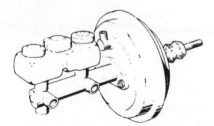

Fig. 9.11. Master cylinder with attached servo and integral reservoir fitted to North American models

15 Master cylinder (single type) - overhaul

1   Obtain a repair kit which is specifically for the master cylinder which has been fitted to your car.
2   With the master cylinder removed from the car, brush and clean away all external dirt.
3   If the car is equipped with remote type servo units, withdraw the flexible bellows and pushrod before extracting the circlip from its groove in the end of the cylinder body.

4   The internal components can now be extracted and kept in strict sequence for refitting. The components will vary in design according to the tyre of master cylinder.

5   Wash all parts in hydraulic fluid or methylated spirit and examine the piston and cylinder bore surfaces for scratches, scoring or 'bright' wear areas. If these are observed, renew the master cylinder complete. If these components are in good order, discard the old seals and fit new ones from the repair kit, using the fingers only to manipulate them into position.

6   Dip the internal components in clean hydraulic fluid before installing them.

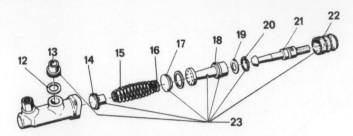

Fig. 9.13. Exploded view of single type master cylinder used in conjunction with a remote servo unit

12	Washer	18	Seal
13	Plug	19	Stop washer
14	Valve	20	Circlip
15	Spring	21	Pushrod
16	Spring plate	22	Boot
17	Collar	23	Items supplied in repair kit

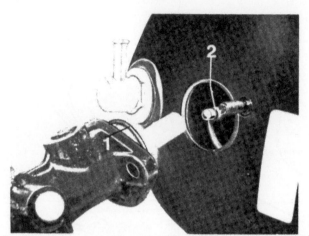

Fig. 9.12. Master cylinder to servo 'O' ring (1) and pushrod shim (2)

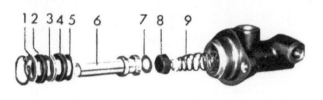

Fig. 9.14. Exploded view of single type master cylinder used in conjunction with attached type servo unit

1	Washer	5	Stop washer
2	Seal	6	Piston
3	Ring	7	Spacer
4	Seal	8	Cup seal
		9	Spring

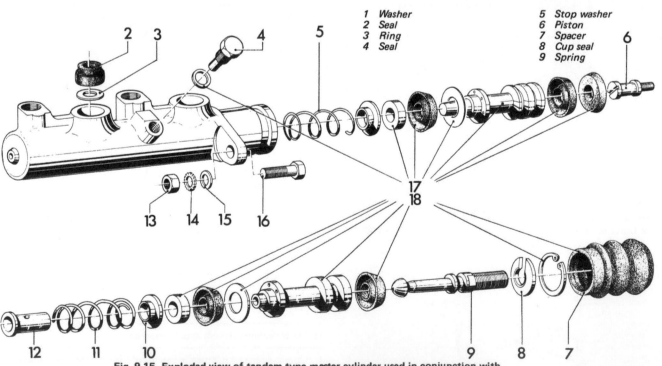

Fig. 9.15. Exploded view of tandem type master cylinder used in conjunction with remote type servo unit

2	Plug	7	Boot	12	Stop screw bush	17	Primary piston components supplied in repair kit
3	Washer	8	Stop washer	13	Mounting bolt component	18	Secondary piston components supplied in repair kit
4	Stop screw	9	Pushrod	14	Mounting bolt component		
5	Primary spring	10	Spring retainer	15	Mounting bolt component		
6	Stop screw	11	Secondary spring	16	Mounting bolt component		

16 Master cylinder (tandem type) - overhaul

1 Obtain a repair kit which is specifically for the master cylinder which has been fitted to your car.
2 With the master cylinder removed from the car, brush and clean away all external dirt.
3 If the car is equipped with remote type servo units, withdraw the flexible bellows and pushrod from the rear end of the master cylinder.
4 Exert a little pressure on the end of the master cylinder piston and unscrew and remove the stop screw.
5 Extract the circlip from the end of the cylinder body and withdraw the secondary piston and allied components.
6 The primary piston and components can be extracted after releasing the special screw. Any difficulty experienced in extracting the internal components can be overcome by applying air from a tyre pump to one of the fluid outlet holes while the others are sealed with plugs.
7 Clean all components in hydraulic fluid or methylated spirit and keep the parts in strict order for refitting.
8 Examine the surfaces of the pistons and cylinder bores for scratches, scoring or 'bright' wear areas. If any are observed, renew the cylinder complete. Where these components are in good condition, discard all rubber seals and obtain the appropriate repair kit. Manipulate the new seals into place using the fingers only. Dip internal components into clean hydraulic fluid before reassembly.

17 Vacuum servo unit - general description

1 A vacuum servo unit is fitted into the brake hydraulic circuit in series with the master cylinder, to provide assistance to the driver when the brake pedal is depressed. This reduces the effort required by the driver to operate the brakes under all braking conditions.
2 The unit operates by vacuum obtained from the induction manifold and comprises basically a booster diaphragm and non-return valve. The servo unit and hydraulic master cylinder are connected together so that the servo unit piston rod acts as the master cylinder pushrod. The driver's braking effort is transmitted through another pushrod to the servo unit piston and its built in control system. The servo unit piston does not fit tightly into the cylinder, but has a strong diaphragm to keep its edges in constant contact with the cylinder wall, so assuring an air tight seal between the two parts. The forward chamber is held under vacuum conditions created in the inlet manifold of the engine and, during periods when the brake pedal is not in use, the controls open a passage to the rear chamber so placing it under vacuum conditions as well. When the brake pedal is depressed, the vacuum passage to the rear chamber is cut off and the chamber opened to atmospheric pressure. The consequent rush of air pushes the servo piston forward in the vacuum chamber and operates the main pushrod to the master cylinder.
3 The controls are designed so that assistance is given under all conditions and, when the brakes are not required, vacuum in the rear chamber is established when the brake pedal is released. All air from the atmosphere entering the rear chamber is passed through a small air filter.
4 Under normal operating conditions the vacuum servo unit is very reliable and does not require overhaul except at very high mileage. In this case it is far better to obtain a service exchange unit, rather than repair the original unit.
5 It is emphasised, that the servo unit assists in reducing the braking effort required at the foot pedal and in the event of its failure, the hydraulic braking system is in no way affected except that the need for higher pedal pressures will be noticed.

18 Vacuum servo filter - renewal

1 At intervals of 30,000 miles (48,000 km) or more frequently in dusty conditions, disconnect the servo pushrod from the brake pedal and pull off the flexible boot.
2 The filter is cut and can be pulled from the pushrod and a new one installed in the same way.
3 Make sure that the slots in the filter and silencer pads are not in alignment and then refit the flexible boot and reconnect the pushrod.

19 Vacuum servo booster unit (pedal connected type) - removal

1 Syphon the fluid from the master cylinder reservoir and disconnect the supply hoses from the master cylinder (not North American versions with integral reservoirs).
2 Disconnect the fluid pressure lines from the master cylinder.
3 Disconnect the vacuum hose from the non-return valve.
4 Remove the clevis pin from the pushrod and then unbolt the servo unit and withdraw it forward.
5 The master cylinder can be unbolted from the front face of the servo unit.
6 Refitting is a reversal of removal but bleed the hydraulic system (Section 23) and check the brake pedal and stop lamp switch setting (Section 25).

20 Vacuum servo booster unit (remote type) - removal and refitting

1 Disconnect the leads from the stop lamp switch (if fitted) and then disconnect the fluid line from the master cylinder at the booster hydraulic cylinder union (photo).
2 Unscrew and remove the hollow bolt which secures the tee-piece to the booster hydraulic cylinder.
3 Disconnect the vacuum hose from the booster and then unbolt the unit from its bracket.
4 Refitting is a reversal of removal but bleed the hydraulic system on completion (Section 23).

21 Flexible brake hoses - inspection, removal and installation

1 Periodically, inspect the condition of the flexible brake hoses. If they appear swollen, chafed or when bent double with the fingers tiny cracks are visible, then they must be renewed.
2 Always uncouple the rigid pipe from the flexible hose first, then release the end of the flexible hose from the support bracket. Now unscrew the flexible hose from the caliper or connector. If this method is followed, no kinking of the hose will occur.
3 When installing the hose, always use a new copper sealing washer.
4 When installation is complete, check that the flexible hose does not rub against the tyre or other adjacent components. Its attitude may be altered to overcome this by releasing its bracket support locknut and twisting the hose in the required direction by not more than one quarter turn.
5 Bleed the hydraulic system.

22 Rigid brake lines - inspection, removal and installation

1 At regular intervals wipe the steel brake pipes clean and examine them for signs of rust or denting caused by flying stones.
2 Examine the fit of the pipes in their insulated securing clips and bend the tongues of the clips if necessary to ensure a positive fit.
3 Check that the pipes are not touching any adjacent components or rubbing against any part of the vehicle. Where this is observed, bend the pipe gently away to clear.
4 Any section of pipe which is rusty or chafed should be renewed. Brake pipes are available to the correct length and fitted with end unions from most BMW dealers and can be made to pattern by many accessory supplies. When installing the new pipes use the old pipes as a guide to bending and do not make any bends sharper than is necessary.
5 The system will of course have to be bled when the circuit has been reconnected.

23 Hydraulic system - bleeding

1 This is not a routine operation and will normally only require to be carried out if a part of the hydraulic system has been disconnected. A spongy feeling in the brake pedal is normally due to air in the system but before bleeding it out, ascertain the reason for its entry. This could be a low fluid level in the reservoir or faulty seals in one of the hydraulic cylinders.

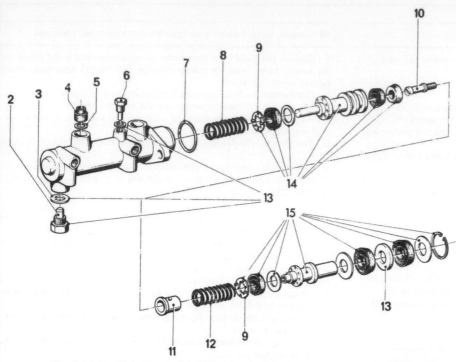

2   Pre-pressure valve
3   Washer
4   Plug
5   Washer
6   Stop screw
7   'O' ring
8   Primary spring
9   Spring retainer
10 Stop screw
11 Stop bush
12 Secondary spring
13 Intermediate ring
14 Primary piston components
    supplied in repair kit
15 Secondary piston components
    supplied in repair kit
16 Cylinder body items
    supplied in repair kit

Fig. 9.16. Exploded view of tandem type master cylinder used in conjunction with attached type servo unit

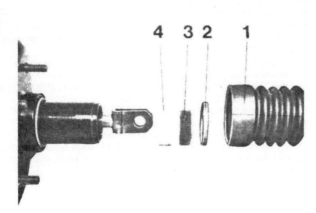

Fig. 9.17. Vacuum servo filter

1   Boot   3   Silencer
2   Retainer   4   Filter

Fig. 9.18. Brake servo vacuum hose and non-return valve

Fig. 9.19. Servo pushrod to pedal connection

1   Clevis pin

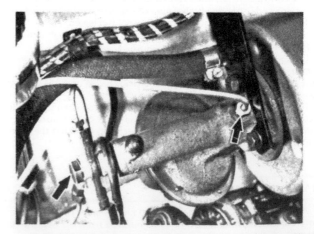

Fig. 9.20. Remote type servo hydraulic cylinder and connections

2   Every two years it is recommended that the fluid is bled from the complete system and new fluid introduced. This is because hydraulic fluid will absorb moisture from the atmosphere which reduces its operating efficiency and can also cause corrosion of the internal components of the system.

3   Gather togther a clean glass jar, a length of flexible tubing which fits tightly over the bleed nipples and a quantity of specified fluid which has been stored in an airtight container and has remained unshaken for 24 hours.

4   Clean round the nipples and remove the dust caps from them.

5   On cars with remote servo units, bleed the hydraulic cylinder attached to the servo unit first and then the rear brakes and the front calipers. On four piston type calipers, bleed from the top outer nipple first followed by the lower inner nipple and then the lower outer one.

6   Push the flexible tube onto the first nipple and submerge the open end of the tube in some fluid poured into the glass jar. The end of the tube must be kept submerged and the master cylinder reservoir topped up throughout the bleeding operations otherwise air will be drawn into the system and the procedure will have to start over again.

7   Open the bleed nipple with a spanner and have an assistant quickly depress the foot pedal. His foot should now be quickly removed from the pedal so that it returns under its own power. Repeat the operation until no more air bubbles can be seen coming from the end of the tube under the fluid in the jar. Tighten the bleed nipple while the foot pedal is fully depressed.

8   Continue the bleeding process in the sequence given in paragraph 5, but always top-up the reservoir with clean fluid before moving on to the next wheel cylinder.

9   Always discard fluid which has been bled from the system or use it only for bleed jar purposes.

---

24 Handbrake cable - renewal

1   Release the handbrake lever fully and remove the flexible gaiter.

2   Release and remove the locknut and adjusting nut from the threaded part of the cable which is to be renewed (see Section 3).

3   Release the cable securing clip from the rear suspension arm.

4   Remove the roadwheel and brake drum and disconnect the handbrake cable from the lever on the trailing brake shoe, as described in Section 5.

5   Pull the cable assembly through its support sleeve on the suspension arm and through the rear brake backplate and remove it.

6   Installation of the new cable is a reversal of removal but adjust it after fitting, as described in Section 3. Make sure that the protective sleeves are correctly located at the suspension arm clip and support sleeve.

---

25 Brake pedal - removal, refitting and adjustment

1   Removal and refitting are carried out after removal of the clutch pedal, as described in Chapter 5, Section 4.

2   On cars with automatic transmission the brake pedal pivot bolt is removed as far clutch models but when refitting, make sure that the longer spacers are placed on the pivot bolt to the left of the pedal arm.

3   When installed, the brake pedal pad should be level with that of the clutch pedal. To achieve this, release the locknut and screw the clevis fork nearest the servo or master cylinder pushrod in or out.

4   Finally check the setting of the brake stop lamp switch, (on some models the switch is installed in the hydraulic circuit). The length of the projecting plunger of the switch (when in contact with the pedal arm - fully released) should be between 0.24 and 0.28 in. (6.0 and 7.0 mm) (photo).

---

26 Vacuum servo non-return valve - renewal

1   Failure of the servo unit to maintain vacuum may be due to a faulty non-return valve (assuming that the connecting hoses are tight and not leaking).

2   One type of valve is screwed directly into the inlet manifold but later models have a valve which is located in the vacuum hose itself.

3   When the later type is renewed, make sure that the directional

arrows or black section of the valve are pointing towards the inlet manifold.

20.1 Location of remote type vacuum servo units used in conjunction with dual circuit braking system

Fig. 9.21. Remote type servo unit mounting bracket

Fig. 9.22. Handbrake cable securing clip and support sleeve

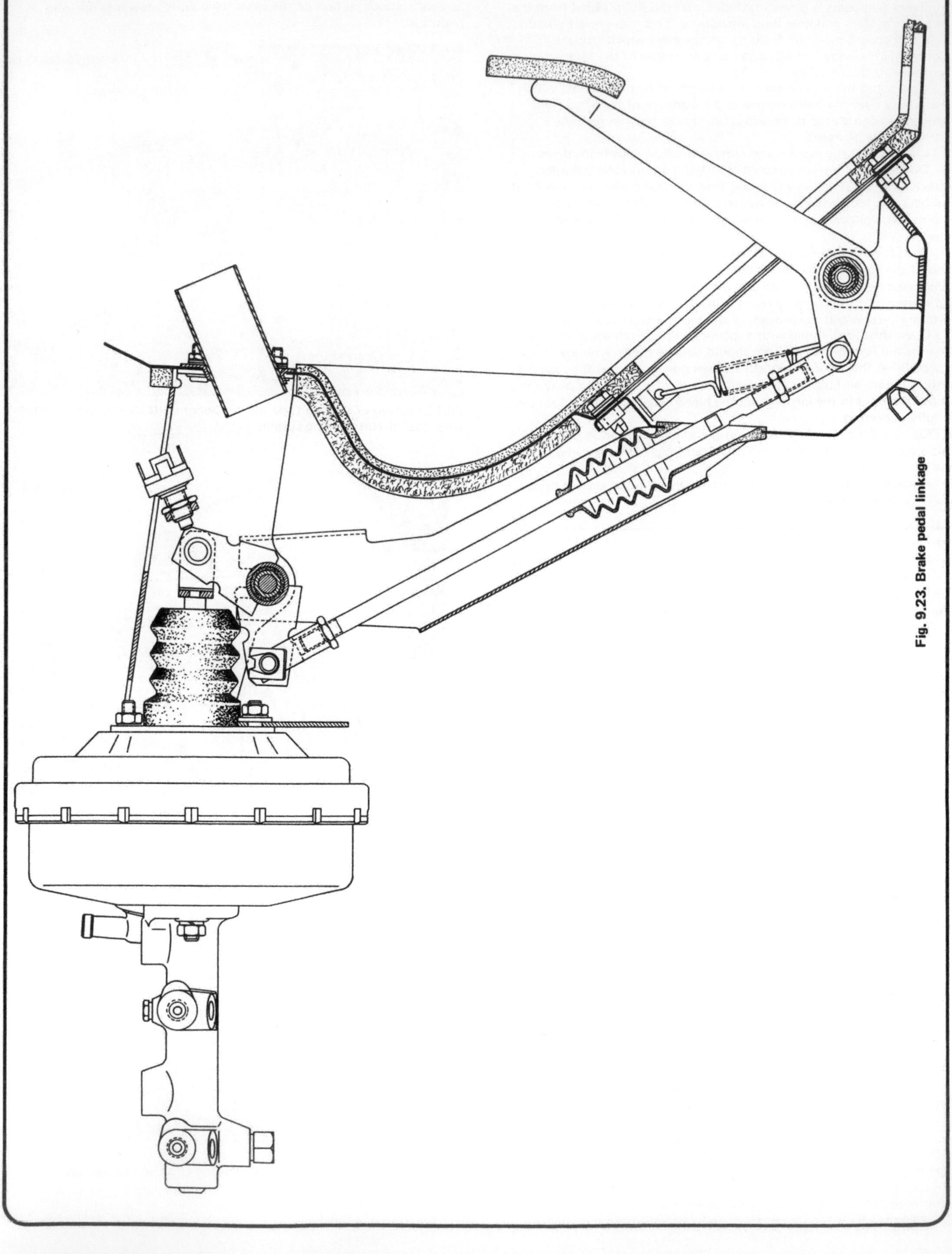

Fig. 9.23. Brake pedal linkage

27 Pressure differential switch and fluid level indicator

1 On some models with a dual circuit hydraulic system, a pressure differential switch is incorporated. This is essentially a switch in which a piston is kept in balance by the equal pressures of the two hydraulic circuits. When the pressure in one circuit drops due to a leak or other cause, the piston is displaced and completes an electrical circuit to illuminate a warning light on the intrument panel.

2 On cars supplied for operation in North America, a reservoir fluid level indicator is installed which operates when the fluid level in one division of the reservoir has fallen low enough for the float to complete an electrical circuit and illuminate a lamp on the instrument panel.

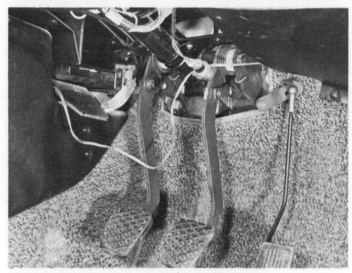

25.4 Pedal arrangement showing brake stop lamp switch

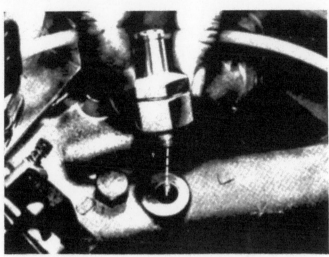

Fig. 9.24. Servo non-return valve screwed directly into manifold

Fig. 9.25. Alternative type non-return valve for servo system

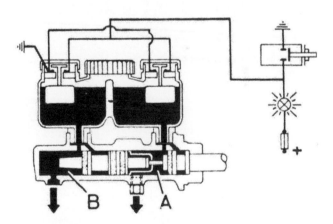

Fig. 9.26. Diagram of reservoir fluid level indicator

A Master cylinder chamber to rear brakes
B Master cylinder chamber to front brakes

28 Fault diagnosis - braking system

Symptom	Reason/s
Pedal travels almost to floorboards before brakes operate	Brake fluid level too low Caliper leaking Master cylinder leaking (bubbles in master cylinder fluid) Brake flexible hose leaking Brake line fractured Brake system unions loose Rear brakes need adjustment
Brake pedal feels springy	New linings not yet bedded-in Brake discs or drums badly worn or cracked Master cylinder securing nuts loose
Brake pedal feels spongy and soggy	Caliper or wheel cylinder leaking Master cylinder leaking (bubbles in master cylinder reservoir) Brake pipe line or flexible hose leaking Unions in brake system loose
Excessive effort required to brake car	Pad or shoe linings badly worn New pads or shoes recently fitted - not yet bedded in Harder linings fitted than standard causing increase in pedal pressure Linings and brake drums contaminated with oil, grease or hydraulic fluid Servo unit inoperative or faulty
Brakes uneven and pulling to one side	Linings and discs or drums contaminated with oil, grease or hydraulic fluid Tyre pressures unequal Radial ply tyres fitted at one end of the car only Brake caliper loose Brake pads or shoes fitted incorrectly Different type of linings fitted at each wheel Anchorages for front suspension or rear suspension loose Brake discs or drums badly worn, cracked or distorted
Brakes tend to bind, drag or lock-on	Air in hydraulic system Wheel cylinders seized Handbrake cables too tight

Chapter 10 Electrical system

Contents

Specifications

System type
1500 and early 1600 ... ...   6V negative earth
All other models ... ...   12V negative earth

Battery
1500 and early 1600 ... ...   77 amp/hr
1502 and 1602 ... ...   36 amp/hr
2000 and 2002 ... ...   44 amp/hr

Dynamo (1500 and early 1600
Type ... ...   Bosch LJ/GEG 200/6/2400R
Cut-in speed ... ...   1600 rpm
Maximum current ... ...   50 amp
Minimum brush length ... ...   0.12 in (5.0 mm)

Alternator

	Later 1600	Early 1602, 2000/2002	Later 1602, 2000 2002 and 1502
Type (Bosch)	35A20	35A20	45A22
Maximum output (watts)	490	490	650
Maximum current (amps)	35	35	45
Charging begins (rpm)	900	900	1150
Minimum brush length	0.28 in (7.0 mm)	0.28 in (7.0 mm)	0.28 in (7.0 mm)

Starter motor

	1500 and early 1600	Later 1600, 1502, 1602	2000, 2002
Type (Bosch)	GF(R)6	EF(R)12	GF(R)12
Output (hp)	0.6	0.8	1.0
Maximum speed (rpm)	—	2400	1300
Current (amps)	—	175	210
Minimum brush length	0.12 in (5 mm)	0.12 in (5 mm)	0.12 in (5 mm)

Voltage regulator (dynamo)
Type ... ...   Bosch RS/VA 200/6
Cut-in voltage ... ...   5.9 to 6.5V
Idling speed voltage ... ...   6.9 to 7.5V
On load voltage ... ...   6.2 to 7.0V
On load current ... ...   65 amps

Voltage regulator (alternator)

Type	Bosch AD1
Operating voltage	14V
Regulator voltage	12V
Output	35 amps at 2700 rpm
Maximum field current	3 amps

Fuses

1500 and early 1600	4 X 8A, 2 X 25A
Later 1600	5 X 8A, 1 X 16A
1602, 2002 (up to 1971)	5 X 8A, 1 X 16A
1502, 2000 Touring and later 1602 and 2002	4 X 5A, 4 X 8A, 4 X 16A
2002 TI	5 X 8A, 1 X 16A
2002 TII	4 X 5A, 4 X 8A, 4 X 16A

Bulbs

	Wattage
Headlamps	45/40
Front parking	4
Front flasher:	
Early 1500/1600	18
All later models	21
Stop lamp:	
Early 1500/1600	18
All later models	21
Reversing lamp:	
Up to 1971	15
Later	21
Tail	5
Rear licence plate	5
Interior:	
2000 Touring	5
Other models	10
Instrument panel and indicator	3
Luggage compartment (2000 Touring only)	10

1 General description

The electrical system is of negative earth type. Some early 1500 and 1600 models had a 6 volt system with dynamo but all later models have a 12 volt system with alternator.

The battery supplies a steady current to the ignition system and for all the electrical accessories. The generator maintains the charge in the battery and the voltage regulator adjusts the charging rate according to the battery's demands. The cut-out prevents the battery discharging to earth through the generator when the engine is switched off and current generation stops.

2 Battery - removal and refitting

1 The battery is located at the front on the left-hand side of the engine compartment (photo).
2 Disconnect the negative terminal first whenever servicing the battery.
3 Then remove the positive terminal, and remove the battery frame holding-down screws and lift the frame away.
4 Lift out the battery carefully to avoid spilling electrolyte on the paintwork.
5 Replacement is a reversal of removal procedure but when reconnecting the terminals, clean off any white deposits present and smear with petroleum jelly.

3 Battery - maintenance and inspection

1 Keep the top of the battery clean by wiping away dirt and moisture.
2 Remove the plugs or lid from the cells and check that the electrolyte level is just above the separator plates. If the level has fallen, add only distilled water until the electrolyte level is just above the separator plates.
3 As well as keeping the terminals clean and covered with petroleum jelly, the top of the battery, and especially the top of the cells, should be kept clean and dry. This helps prevent corrosion and ensures that the battery does not become partially discharged by leakage through

dampness and dirt.
4 Once every three months, remove the battery and inspect the battery securing bolts, the battery clamp plate, tray and battery leads for corrosion (white fluffy deposits on the metal which are brittle to touch). If any corrosion is found, clean off the deposits with ammonia and paint over the clean metal with an anti-rust/anti-acid paint.
5 At the same time inspect the battery case for cracks. If a crack is found, clean and plug it with one of the proprietary compounds marketed for this purpose. If leakage through the crack has been excessive then it will be necessary to refill the appropriate cell with fresh electrolyte as detailed later. Cracks are frequently caused to the top of the battery cases by pouring in distilled water in the middle of winter *after* instead of *before* a run. This gives the water no chance to mix with the electrolyte and so the former freezes and splits the battery case.
6 If topping-up the battery becomes excessive and the case has been inspected for cracks that could cause leakage, but none are found, the battery is being over-charged and the voltage regulator will have to be checked and reset.
7 The specific gravity of a fully charged battery when tested with a hydrometer at an electrolyte temperature of 68°F (20°C) should be 1.260 with a variation between cells not exceeding 0.025.

4 Electrolyte replenishment

1 If the battery is in a fully charged state and one of the cells maintains a specific gravity reading which is .025 or more lower than the others, and a check of each cell has been made with a voltage meter to check for short circuits (a four to seven second test should give a steady reading of between 1.2 to 1.8 volts), then it is likely that electrolyte has been lost from the cell with the low reading at some time.
2 Top-up the cell with a solution of 1 part sulphuric acid to 2.5 parts of water. If the cell is already fully topped up draw some electrolyte out of it with a pipette.
3 When mixing the sulphuric acid and water **never add water to sulphuric acid** - always pour the acid slowly onto the water in a glass container. **If water is added to sulphuric acid it will explode.**
4 Continue to top-up the cell with the freshly made electrolyte and then recharge the battery and check the hydrometer readings.

5 Battery charging

1 In winter time when heavy demand is placed upon the battery, such as when starting from cold, and much electrical equipment is continually in use, it is a good idea to occasionally have the battery fully charged from an external source at the rate of 3.5 or 4 amps.

2 Continue to charge the battery at this rate until no further rise in specific gravity is noted over a four hour period.

3 Alternatively, a trickle charger charging at the rate of 1.5 amps can be safely used overnight.

4 Specially rapid 'boost' charges which are claimed to restore the power of the battery in 1 to 2 hours are most dangerous as they can cause serious damage to the battery plates.

5 Take extreme care when making circuit connections to a vehicle fitted with an alternator and observe the following. When making connections to the alternator from a battery charger always match correct polarity. Before using electric-arc welding equipment to repair any part of the vehicle, disconnect the connector from the alternator and disconnect the positive battery terminal. Never start the car with a battery charger connected. Always disconnect both battery leads before using a mains charger. If boosting from another battery, always connect in parallel using heavy cable.

2.1 Location of battery

6 Dynamo - removal, overhaul and refitting

1 Disconnect the lead from the battery negative terminal.

2 Identify the dynamo connecting cables and remove them from the dynamo terminals.

3 Slacken the dynamo mounting bolts and adjustment strap bolt, push the dynamo in towards the engine and slip the driving belt from its pulley.

4 Remove the mounting and adjustment strap bolts and lift the dynamo from its mounting brackets.

5 Remove the pulley retaining nut. To prevent the pulley rotating during this operation, place an old belt in the pulley groove and grip the belt as close to the pulley as possible in the jaws of a vice.

6 Release the cover band screw and slip the cover band to the rear.

7 Prise back the brush springs and lift the brushes from their holders. If the brushes are worn to their minimum specified length, renew them.

8 Unscrew the two tie-bolts and withdraw the armature assembly from the yoke.

9 If the commutator is black and discoloured, clean it with fuel and polish it with fine glass paper/not emery cloth). If it is scored or damaged, it can be refinished by your auto-electrical company.

10 If the segment insulators are flush with the surface of the segments then they can be carefully undercut using a ground down hacksaw blade to give a square cornered cut to the groove.

11 Testing of the armature and field coils is best left to a qualified auto-electrician as special equipment is needed. It is doubtful whether in the event of a fault being found in either of these parts the repair of a dynamo would be economical and the installation of a good second-hand or reconditioned unit is to be recommended instead.

12 Reassembly and installation are reversals of dismantling but adjust the drivebelt tension so that it can be depressed by between 0.20 and 0.40 in (5.0 and 10.0 mm) at the centre of its top run.

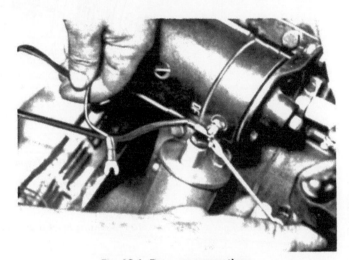

Fig. 10.1. Dynamo connections

7 Voltage regulator (dynamo) - removal and refitting

1 Disconnect the lead from the battery negative terminal.

2 Disconnect the leads from the top terminals of the voltage regulator which is located on the bulkhead within the engine compartment. Identify the leads for exact replacement to their original terminals.

3 Unbolt the regulator and lift it far enough away to be able to disconnect the leads from its lower terminals.

4 If the regulator unit is faulty (this can best be tested by your auto-electrical company) renew it with one of the same type.

5 Refitting is a reversal of removal.

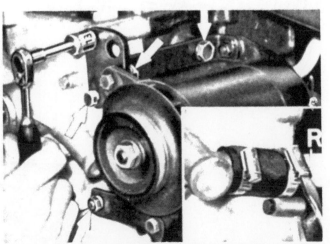

Fig. 10.2. Dynamo mounting bolts (Inset) coolant hose at rear of dynamo

Fig. 10.3. Removing dynamo cover band

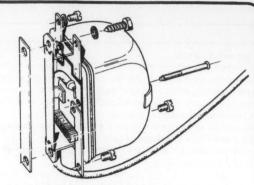

Fig. 10.6. Voltage regulator (dynamo type)

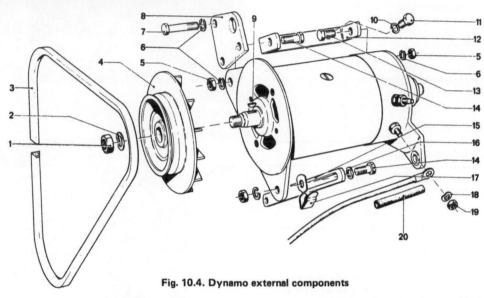

Fig. 10.4. Dynamo external components

1	Pulley nut	6	Front mounting	11	Mounting bracket assembly	16	Washer
2	Lockwasher	7	Front mounting	12	Mounting bracket assembly	17	Cable clip
3	Drivebelt	8	Front mounting	13	Mounting bracket assembly	18	Lockwasher
4	Pulley	9	Key	14	Mounting bracket assembly	19	Nut
5	Front mounting	10	Mounting bracket assembly	15	Adjustment strap	20	Insulator

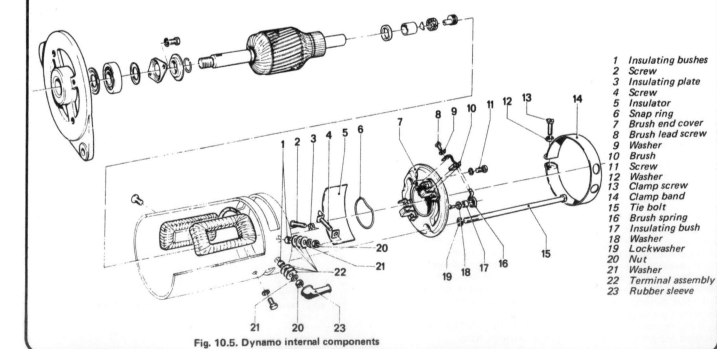

1	Insulating bushes
2	Screw
3	Insulating plate
4	Screw
5	Insulator
6	Snap ring
7	Brush end cover
8	Brush lead screw
9	Washer
10	Brush
11	Screw
12	Washer
13	Clamp screw
14	Clamp band
15	Tie bolt
16	Brush spring
17	Insulating bush
18	Washer
19	Lockwasher
20	Nut
21	Washer
22	Terminal assembly
23	Rubber sleeve

Fig. 10.5. Dynamo internal components

8 Alternator - removal and refitting

1   Disconnect the leads from the battery terminals (never disconnect a lead from the battery while the engine is running).
2   Disconnect the cables and the multi pin plug from the rear face of the alternator.
3   *On fuel injection models,* remove the battery from the engine compartment, also the stabiliser bar (see Chapter 11).
4   Release the mounting and adjustment strap bolts and push the alternator in towards the engine and slip the drivebelt from the alternator pulley.
5   Remove the mounting and adjustment strap bolts and lift the alternator from its mounting brackets (photo).
6   Refitting is a reversal of removal but adjust the tension of the drivebelt, as described in Section 6, paragraph 12.

9 Alternator and regulator - testing

1   If the ignition warning lamp remains illuminated when the engine is running and it is confirmed that the drivebelt has not broken and it is correctly tensioned, carry out the following test.
2   Stop the engine and pull the multi-pin plug from the bottom of the regulator unit.
3   Using a piece of wire bridge the blue and black cable pins of the plug.
4   Start the engine and run it at a speed of about 1000 rpm. If the ignition warning lamp goes off immediately, then the voltage regulator is defective. If the lamp remains on or glows even dimly, the alternator is at fault.
5   A faulty regulator should be renewed. The alternator should only be overhauled to a limit of installing new brushes otherwise it should be repaired by your BMW agent or a new or reconditioned unit obtained. This arrangement is recommended due to the need for special tools and equipment also the fragile nature of the diodes incorporated in the alternator.

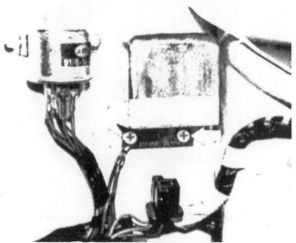

Fig. 10.8. Alternator regulator and connecting plug

Fig. 10.9. Bridging contacts of alternator regulator plug

Fig. 10.7. Alternator terminals

8.5 Removing alternator

10 Alternator - brush renewal

1   Remove the three screws from the drive end plate.
2   Withdraw the rotor assembly.
3   Withdraw the diode carrier far enough to be able to pull off the cable connectors and then unbolt the carbon brush carrier.
4   Remove the worn brushes by unsoldering their leads.
5   Install the new brushes and resolder their leads. Do not let solder flow down the braided copper wire but solder only the tip of the lead and restrict the heat of the soldering iron as much as possible.
6   Push the brushes into their holders and locate the brush springs so that they retain the brushes well up in their holders.
7   Clean the slip rings with a fuel moistened cloth and then reassemble the alternator making sure that the wave washer is located on the front of the rotor shaft bearing and that the diode carrier is towards the top of the alternator.
8   Release the brush springs so that the brushes bear on the slip rings. This can be done by inserting a probe into the multi-pin socket of the alternator.

Fig. 10.10. Alternator drive end plate screws

Fig. 10.11. Withdrawing alternator rotor assembly

Fig. 10.12. Withdrawing alternator diode carrier

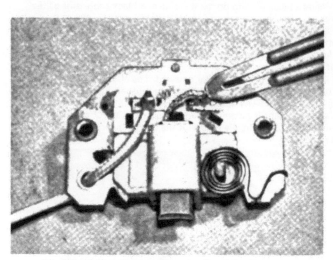

Fig. 10.13. Unsoldering alternator brush lead

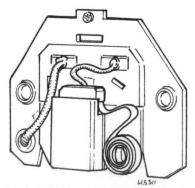

Fig. 10.14. Brushes retained in retracted position by springs

Fig. 10.15. Releasing brush springs after assembly of rotor

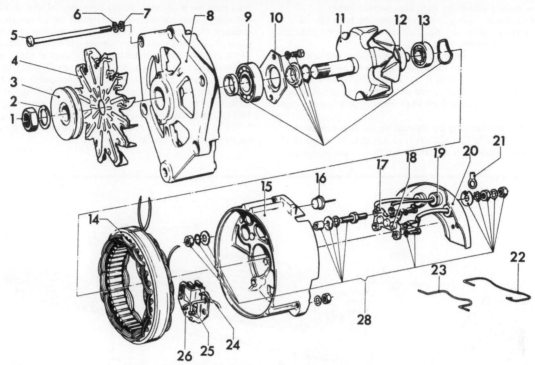

Fig. 10.16. Exploded view of the alternator

1	Pulley retaining nut	8	Drive end cover
2	Lockwasher	9	Bearing
3	Pulley	10	Bearing retainer
4	Fan	11	Rotor
5	Tie bolt	12	Slip ring
6	Washer	13	Bearing
7	Lockwasher	14	Stator

15	Rear cover	22	Plug securing spring
16	Negative diode	23	Plug securing spring
17	Diode plate	24	Brushes
18	Exciter diode	25	Brush spring
19	Positive diode	26	Brush holder
20	Diode carrier	27	Supplied as repair kit
21	Terminal tag	28	Supplied as repair kit

11 Starter motor - description, removal and installation

1   This type of starter motor incorporates a solenoid mounted on the starter motor body. When the ignition switch is operated, the solenoid moves the starter drive pinion, through the medium of the shift lever, into engagement with the flywheel starter ring gear. As the solenoid reaches the end of its stroke and with the pinion by now fully engaged with the flywheel ring gear, the fixed and moving contacts close and energise the starter motor to rotate the engine.

This fractional pre-engagement of the starter drive does much to reduce the wear on the flywheel ring gear associated with inertia type starter motors.

2   To remove the starter motor, first disconnect the lead from the battery negative terminal.

3   Disconnect the leads from the starter motor solenoid terminals. Release the starter motor front mounting (photos).

4   Unbolt the starter motor from the bellhousing and withdraw it (photo).

5   Installation is a reversal of removal.

12 Starter motor - overhaul

1   Servicing operations should be limited to renewal of brushes, renewal of the solenoid, the overhaul of the starter drive gear and cleaning the commutator.

2   The major components of the starter should normally last the life of the unit and in the event of failure, a factory exchange replacement should be obtained.

3   Disconnect the motor field winding lead from the solenoid terminal.

4   Remove the screws which secure the solenoid to the drive end cover and withdraw the solenoid at the same time unhooking it from drive engagement lever.

5   Remove the dust cap from the end of the starter motor and extract the lockwasher, shims and gasket.

6   Unscrew the tie bolts and withdraw the motor end cover.

7   Extract the carbon brushes from their holders and remove the brush mounting plate.

8   Withdraw the yoke from the drive end cover and then unscrew the engagement lever pivot bolt.

11.3a Starter motor leads

11.3b Starter motor front mounting bracket

11.4 Removing the starter motor

9 Withdraw the armature complete with engagement lever.
10 Using a piece of suitable tubing, drive the stop ring back up the armature shaft to expose the jump ring. Extract the jump ring and pull off the starter drive components.
11 Measure the overall length of each of the two brushes and where they are worn below the minimum recommended (see Specifications) renew them by unsoldering and resoldering. Ensure that each brush slides freely in its holder. If necessary, rub with a fine file and clean any accumulated carbon dust or grease from the holder with a fuel moistened rag.
12 Normally, the commutator may be cleaned by holding a piece of non-fluffy rag moistened with fuel against it as it is rotated by hand. If on inspection, the mica separators are level with the copper segments

then they must be undercut by between 0.020 and 0.032 in (0.5 to 0.8 mm). Undercut the mica separators of the commutator using an old hacksaw blade ground to suit. The commutator may be polished with a piece of very fine glass paper - never use emery cloth as the carborundum particles will become embedded in the copper surfaces.
13 Wash the components of the drive gear in paraffin and inspect for wear or damage, particularly to the pinion teeth and renew as appropriate. Refitting is a reversal of dismantling. Oil the sliding surfaces of the pinion assembly with a light oil, applied sparingly.
14 Note that the endfloat of the armature must be between 0.004 and 0.006 in (0.1 and 0.15 mm). This can be adjusted by varying the shims located under the starter motor dust cap.

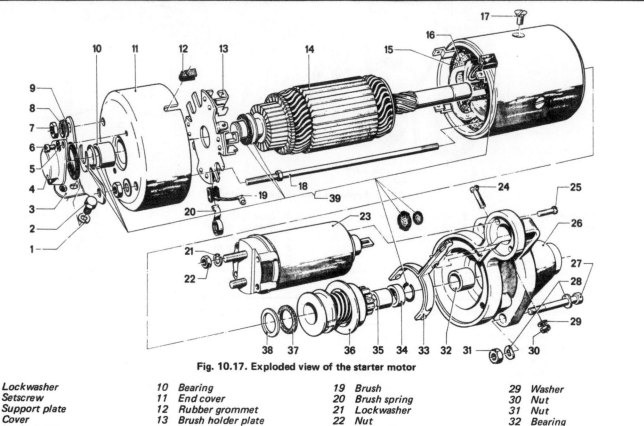

Fig. 10.17. Exploded view of the starter motor

1	Lockwasher	10	Bearing	19	Brush	29	Washer
2	Setscrew	11	End cover	20	Brush spring	30	Nut
3	Support plate	12	Rubber grommet	21	Lockwasher	31	Nut
4	Cover	13	Brush holder plate	22	Nut	32	Bearing
5	Lockwasher	14	Armature	23	Solenoid	33	Engagement lever
6	Screw	15	Insulator	24	Engagement lever pivot bolt	34	Stop ring
7	Nut	16	Field coils	25	Solenoid securing screw	35	Jump ring
8	Lockwasher	17	Screw	26	Drive end cover	36	Starter drive/clutch assembly
9	Washer	18	Tie bolt	27	Starter securing bolt	37	Packing
				28	Lockwashers	38	Shim

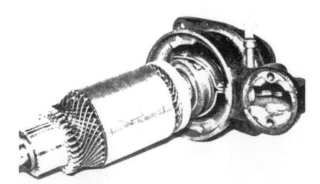

Fig. 10.18. Removing starter motor engagement lever pivot bolt

Fig. 10.19. Starter drive stop ring (1) and jump ring (2)

13 Fuses

1 The fuse box is located under the bonnet on the wheel arch.
2 The number of circuits protected and the fuse sequence varies according to model but they are tabulated on a sticker adjacent to the fusebox.
3 Fuses may be of 5, 8 or 16 amp capacity according to the circuit protected and if one has blown, this can be seen through the transparent lid of the fusebox.
4 Always renew a fuse with one of similar rating and if it blows repeatedly, establish the cause which is often a short circuit due to faulty insulation of the wiring.
5 The radio is protected by a separate in-line fuse in the power feed cable.

13.1 Fuse box

14 Direction indicator/hazard warning flasher relay - testing and renewal

1 If the direction indicators do not operate, carry out the following checks:
    (i)   *Inspect for blown fuse.*
    (ii) *Test security of all leads and connections.*
    (iii) *Inspect switch mechanism.*
2 If the indicator lamps flash too slowly or too quickly or the fascia indicator lamp does not go out, check the lamp units for a burnt out bulb and also for a loose connection. If the flashing cycle is irregular, check for a bulb of incorrect wattage.
3 Where the bulbs, switch and wiring are found to be in order, the flasher unit itself must be at fault.
4 Access to the flasher relay is obtained by removing the outer trim panel from below the instrument panel.
5 To test that current is reaching the relay, connect a test lamp between terminal '+49' and earth and switch on the ignition. If the lamp illuminates, renew the relay.
6 A more extensive test can be carried out if the relay is removed and a lead connected between the battery positive terminal and '+49' terminal of the relay while an earth return is connected between the body and terminal '−31' of the relay. Now connect two test lamps, each of 21 watts to terminal '49a' while a 3 watt bulb is connected to terminal 'C' of the relay. If the flasher unit is in good order, the lamps should flash at regular intervals.

15 Windscreen wiper delay relay - testing and renewal

1 This relay (when fitted) is located adjacent to the ignition coil on the engine compartment rear bulkhead.
2 To test for a fault first switch on the ignition and connect a test lamp between each of the relay terminals '+15' and '31b' in turn and earth. If the lamp illuminates, the relay is faulty and must be renewed.
3 A more extensive test can be carried out by removing the relay and connecting a lead between the battery positive terminal and terminal '+15' of the relay and an earth lead between the body and terminal '31b' of the relay. Connect a test lamp between terminal 'M' of the relay and the body when the lamp should illuminate for a period of five seconds, extinguish and then illuminate again for a similar period and keep repeating this cycle.

Fig. 10.20. Direction indicator/hazard warning flasher relay

Fig. 10.21. Windscreen wiper delay relay

16 Bulbs - renewal

1 Always renew a bulb with one of similar rating and type.

Headlamps (bulb type) and parking lamp
2 Open the bonnet and remove the protective cover from the rear of the lamp (photo).
3 Withdraw the bulb holder and remove the bayonet type bulb (photo).
4 The parking lamp bulbs are held in the headlamp reflector by a spring.
5 When refitting the bulb holder note the small alignment tag and recess to ensure correct installation.

Headlamps (sealed beam type)
6 Open the bonnet and remove the protective cover from the rear of the lamp.
7 Pull the plug from the rear of the lamp.
8 Remove the radiator grille.
9 Unscrew and remove the headlamp sealed beam retaining ring screws and withdraw the retaining ring and lamp unit.
10 Installation of the sealed beam unit is a reversal of removal. Provided the adjustment screws are not altered, the beam alignment will not be altered.

Rear lamps
11 Raise the lid of the luggage boot and remove the two knurled nuts which retain the rear lamp cluster. Withdraw the cluster and renew the bulbs as necessary (photos).

Rear number plate lamp
12 Remove the lens/frame assembly from the rear bumper (two screws) and extract the festoon type bulb.

Interior lamp
13 Prise the lamp from its recess and extract the festoon type bulb (photo).

Front direction indicator lamps
14 Remove the lens/frame assembly (two screws) and extract the bayonet fixing type bulb (photo).

Instrument panel and warning lamp bulbs
15 Access to some of these bulbs is gained by removing the padded cover at the base of the instrument panel and reaching up behind the panel. The bulbs are of wedge-base type and can be released after twisting their plastic holders from their recesses. Refer also to Section 21.

Automatic transmission speed selector indicator lamp
16 Pull off the plastic cover from the selector indicator and extract the two bulbs.

---

17 Headlamps - removal and refitting

1 To remove a headlamp assembly, first withdraw the radiator grille.
2 Withdraw the bulb holder or sealed beam unit according to type, as previously explained.
3 Unscrew the four securing nuts from the rear of the lamp mounting flange and withdraw the lamp assembly forwards.
4 Installation is a reversal of removal but adjust the beams on completion of the work.

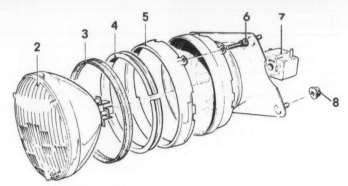

Fig. 10.22. Sealed beam type headlamp components

2 *Sealed beam unit*   6 *Adjusting screw*
3 *Gasket*   7 *Connecting plug*
4 *Spacer*   8 *Mounting nut*
5 *Retaining ring*

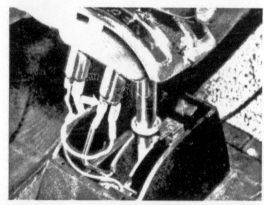

Fig. 10.23. Selector (automatic transmission) indicator lamps

16.2 Removing a headlamp rear cover

16.3 Removing a headlamp bulb holder

16.11a Rear lamp cluster and a securing knob

16.11b Rear lamp cluster removed

16.13 Interior lamp and bulb

16.14 Front flasher lamp

18 Headlamp - beam alignment

1 It is recommended that headlamp beams are always aligned using optical setting equipment at your service station.

2 In an emergency the light pattern may be altered by opening the bonnet and turning the plastic knobs of the adjustment screws at the rear of the headlamp. The lower screw controls horizontal adjustment and the top screw the vertical adjustment.

19 Steering column switches - removal and refitting

1 Disconnect the lead from the battery negative terminal.

2 *On early models,* remove the steering column lower shroud and the lower section of the instrument panel trim. Remove the steering wheel (Chapter 11) and the steering column upper shroud. Disconnect the switch multi-pin plugs and lift out the horn contact ring. The switch screws can now be removed and the switches drawn upwards off the steering column.

3 *On later models,* the steering wheel need not be removed as a thin screwdriver can be inserted through the holes in the steering wheel hub and the switch screws removed and the switches detached from their support plate (photos).

4 When refitting the dipperswitch, this is a reversal of removal but when installing the direction indicator switch, the clearance between the dog on the switch and column reset cam must be 0.012 in (0.3 mm).

20 Ignition switch - removal and refitting

1 Disconnect the lead from the battery negative terminal.

2 Remove the steering column lower shroud and the centre section of the fascia trim (photo).

3 Turn the ignition key to the 'HALT' position.

4 Remove the two screws which secure the ignition switch to the steering column lock assembly. Before disconnecting the cables from the switch terminals, mark them for correct refitting.

5 When refitting the switch, make sure that the lug on the switch is

in engagement with the slot in the steering column lock.

6 Removal and installation of the steering column lock is described in Chapter 11.

21 Instrument panel and instruments - removal and refitting

1 Disconnect the lead from the battery negative terminal.

2 Remove the steering column lower shroud and the trim panel from below the instrument panel.

3 Reach up behind the instrument panel and disconnect the speedometer drive cable from the speedometer and unscrew the two panel securing knurled nuts.

4 Remove the securing screws from the front face of the instrument panel and from the anti-glare shroud over the instruments.

5 Pull the panel far enough forward to be able to disconnect the multi-pin plugs and leads.

6 Withdraw the panel and detach the individual instruments as required. The panel lamps are also accessible for removal or refitting, also the panel-mounted switches (photos).

7 Refitting of the instrument panel is a reversal of removal.

22 Windscreen wiper blades and arms - removal and refitting

1 It is recommended that the wiper blades are renewed every two years or when they fail to clean the screen effectively.

2 To remove a blade, pull the arm outwards from the screen so that it is parallel to the bonnet. It will lock in this position.

3 Depress the locking tab and pull the blade from the arm (photo).

4 The wiper arm can be removed (complete with blade) from the driving spindle if the small spring retainer which locks it to the spindle is first pulled aside with a small screwdriver (photo).

5 Refitting is a reversal of removal but do not push the arm fully onto the splines of the driving spindle before the alignment of the arm with the base of the windscreen is checked. If required, the position of the arm can be altered slightly by withdrawing it and moving it a spline, or two, in either direction.

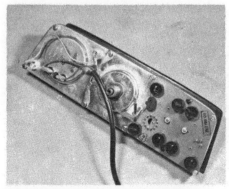
19.3a Removing a steering column switch (late model)

19.3b Side view of steering column switches

20.2 Removing a steering column shroud screw

21.6a Reverse side of instrument panel

21.6b Instruments separated from panel

21.6c Instrument panel removed to provide access to cigar lighter and other switches

Fig. 10.24. Setting diagram for direction indicator switch

A = 0.012 (0.3 mm)

Fig. 10.25. Speedometer cable connection and knurled securing nuts at rear of instrument panel

23 Windscreen wiper motor and linkage - removal and installation

1 Open the bonnet and support it in its fully open position.
2 Check that the motor has been switched off by the wiper switch and not by turning the ignition key otherwise the motor crank will not be in its parked position (photo).
3 Mark the relative position of the crank arm to the motor driveshaft and then unscrew the retaining nut and disconnect the arm from the shaft.
4 Disconnect the motor leads at the connector and detach the separate earth cable.
5 Unscrew and remove the mounting bolts and withdraw the motor from its cavity in the engine compartment rear bulkhead.
6 The linkage can be withdrawn if the wiper arms are first removed and the driving spindle nuts released.
7 Refitting is a reversal of removal.

24 Radio - removal and refitting

1 The radio installed as original equipment is housed in the central console.
2 Remove the screws which secure the tray of the console and lift the tray upwards, easing it from the gearshift lever boot and over the gear lever knob.
3 Extract the screws from the now exposed console brackets and pull the console far enough to the rear to be able to disconnect the radio earth, feed and aerial leads. The radio can then be detached from the console front section.
4 Refitting is a reversal of removal.

25 Radio (aftermarket) - installation

1 If a radio is to be installed in a car not previously fitted with one, the following guide lines are given but are not intended as comprehensive fitting instructions. It is suggested that the radio and loudspeaker are installed in the front section of the centre console and that the aerial is located on the left-hand windscreen pillar.

   a) Check the radio for compatible polarity (negative earth).
   b) Follow the radio manufacturer's instructions.
   c) Always fit an in-line fuse in the radio feed cable.
   d) The ignition system will already be suppressed but additional interference suppressors will almost certainly be required to be fitted to the coil, the voltage regulator and the alternator. Where it is difficult to eliminate all interference, an earth bonding strap may be required to be fitted between the bonnet lid and the body and even contact strips between the roadwheels and hubs. Before installing all these items, fit one and observe the result.

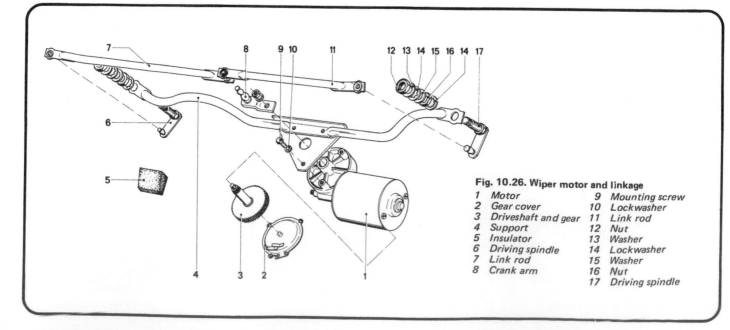

Fig. 10.26. Wiper motor and linkage

1	Motor	9	Mounting screw
2	Gear cover	10	Lockwasher
3	Driveshaft and gear	11	Link rod
4	Support	12	Nut
5	Insulator	13	Washer
6	Driving spindle	14	Lockwasher
7	Link rod	15	Washer
8	Crank arm	16	Nut
		17	Driving spindle

22.3 Removing a windscreen wiper blade

22.4 Removing a windscreen wiper arm

23.2 Location of windscreen wiper motor at rear of engine compartment

Fig. 10.27. Removing tray from centre console

Fig. 10.28. Location of radio

26 Fault diagnosis - electrical system

Symptom	Reason/s
Starter motor fails to turn engine No electricity at starter motor	Battery discharged Battery defective internally Battery terminal leads loose or earth lead not securely attached to body Loose or broken connections in starter motor circuit Starter motor switch or solenoid faulty
Electricity at starter motor: faulty motor	Starter brushes badly worn, sticking, or brush wires loose Commutator dirty, worn or burnt Starter motor armature faulty Field coils earthed
Starter motor turns engine very slowly Electrical defects	Battery in discharged condition Starter brushes badly worn, sticking, or brush wires loose Loose wires in starter motor circuit
Starter motor operates without turning engine Mechanical damage	Pinion or flywheel gear teeth broken or worn
Starter motor noisy or excessively rough engagement Lack of attention or mechanical damage	Pinion or flywheel gear teeth broken or worn Starter motor retaining bolts loose
Battery will not hold charge for more than a few days Wear or damage	Battery defective internally Electrolyte level too low or electrolyte too weak due to leakage Plate separators no longer fully effective Battery plates severely sulphated
Insufficient current flow to keep battery charged	Battery plates severely sulphated Fan belt slipping Battery terminal connections loose or corroded Alternator not charging Short in lighting circuit causing continual battery drain Regulator unit not working correctly
Ignition light fails to go out, battery runs flat in a few days Alternator not charging	Fan belt loose and slipping or broken Brushes worn, sticking, broken or dirty Brush springs weak or broken Commutator dirty, greasy, worn or burnt Alternator field coils burnt, open or shorted Commutator worn Pole pieces very loose
Regulator or cut-out fails to work correctly	Regulator incorrectly set Cut-out incorrectly set Open circuit in wiring of cut-out and regulator unit
Horn Horn operates all the time	Horn push either earthed or stuck down Horn cable to horn push earthed
Horn fails to operate	Blown fuse Cable or cable connection loose, broken or disconnected Horn has an internal fault
Horn emits intermittent or unsatisfactory noise	Cable connections loose Horn incorrectly adjusted
Lights Lights do not come on	If engine not running, battery discharged Sealed beam filament burnt out or bulbs broken Wire connections loose, disconnected or broken Light switch shorting or otherwise faulty

Lights come on but fade out	If engine not running battery discharged Light bulb filament burnt out or bulbs or sealed beam units broken Wire connections loose, disconnected or broken Light switch shorting or otherwise faulty
Lights give very poor illumination	Lamp glasses dirty Lamp badly out of adjustment
Lights work erratically - flashing on and off, especially over bumps	Battery terminals or earth connection loose Light not earthing properly Contacts in light switch faulty

Wipers

Wiper motor fails to work	Blown fuse Wire connections loose, disconnected, or broken Brushes badly worn Armature worn or faulty Field coils faulty
Wiper motor works very slowly and takes excessive current	Commutator dirty, greasy or burnt Armature bearings dirty or unaligned Armature badly worn or faulty
Wiper motor works slowly and takes little current	Brushes badly worn Commutator dirty, greasy or burnt Armature badly worn or faulty
Wiper motor works but wiper blades remain static	Wiper motor gearbox parts badly worn
Wipers do not stop when switched off or stop in wrong place	Auto-stop device faulty

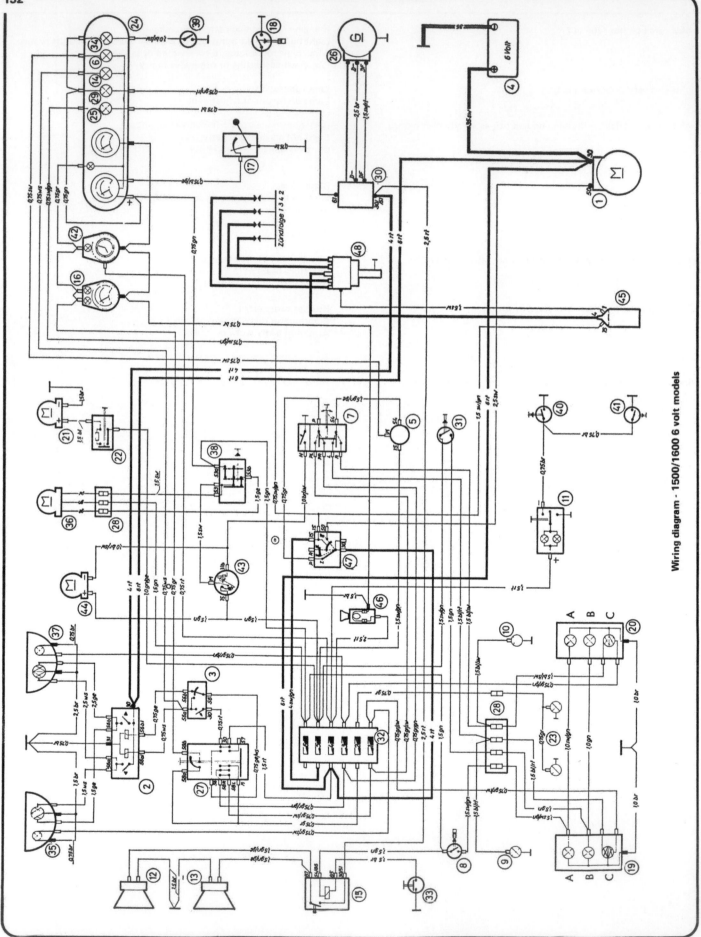

Wiring diagram - 1500/1600 6 volt models

Wiring diagram BMW 1500/1600

1 Starter
2 Dip relay
3 Dipswitch
4 Battery
5 Flasher unit
6 Flasher telltale
7 Flasher/parking light switch with washer contact
8 Stop light switch
9 Turn indicator front LH
10 Turn indicator front RH
11 Roof light
12 Two-tone horn 1
13 Two-tone horn 2
14 Main beam telltale
15 Horn relay
16 Speedometer
17 Fuel gauge mechanism
18 Oil pressure contact
19 Rear light LH
A Stop light
B Reversing light
C Turn indicator/parking light rear
21 Heater blower motor
22 Heater switch

23 Number plate light
24 Combined instrument
25 Battery charge telltale
26 Generator
27 Light switch
28 Cable connector
29 Oil pressure telltale
30 Regulator
31 Reversing light switch
32 Fuse box
33 Horn button
34 Choke telltale
35 Headlight LH
36 Screenwiper motor
37 Headlight RH
38 Screenwiper motor
39 Choke cable contact
40 Door switch LH
41 Door switch RH
42 Clock
43 Delay relay
44 Screenwasher pump
45 Coil
46 Cigar lighter
47 Ignition/starter switch
48 Distributor

Cable coding

0.75 sq mm cross-section

Colour:

BL = blue
BR = brown
GE = yellow
GN = green
GR = grey
RT = red
SW = black
WS = white

Fuses

1—2—3—5 = 8 amp
4—6 = 25 amp

Ignition/starter switch:

Positions:

I   Halt
II   Garage
III   Drive
IV   Start

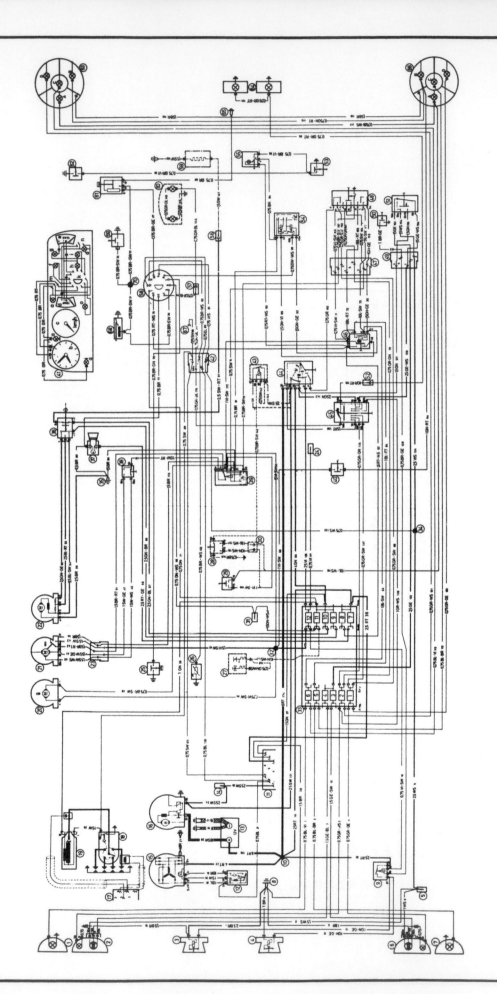

Wiring diagram - 1602 (typical)

Wiring diagram BMW 1602

1 Front right-hand flasher
2 RH headlight with parking light
3 RH fanfare
4 Left-hand fanfare
5 Connection for fog lamp relay
6 LH headlight with parking light
7 Front left-hand flasher
8 Earth (ground)
9 Fanfare relay
10 Solder point
11 Test equipment connection
12 Regulator
13 Battery
14 Electrical system connection
15 Generator
16 Starter
17 Connection for test equipment with lead and pick up
18 Distributor
19 Ignition coil
20 Windshield washer pump
21 Windshield wiper motor
22 Blower motor
23 5 pin plug terminal to wiper motor
24 Radio connection
25 Oil pressure switch
26 Remote reading therm. transmitter
27 Auto. starter carburettor (only with automatic transmission)
28 Reversing light switch with starter lock (only with automatic transmission)
29 Reversing light switch
30 2 pin plug term. (only for automatic trans.)
31 Connection for fuel pump
32 Solder point
33 Fuse box
34 Solder point
35 Earth (ground)
36 Blower switch
37 Cigar lighter
38 Wiper speed switch
39 Wiper-washer transmitter
40 Brake light switch
41 Instrument cluster
   a) Illuminated scale
   b) Clock
   c) Speedometer
   d) Coolant temperature gauge
   e) Fuel gauge
   f) Charging lamp (red)
   g) Oil pressure control (orange)
   h) High beam telltale (blue)
   i) Flasher (green)
   k) 12 pin plug terminal
   m) 3 pin plug (clock)
   n) 3 pin plug (rev. counter)
   o) Central control lamp (choke, handbrake, fuel)
42 5 pin plug terminal
43 Starter relay (only with automatic trans.)
44 Ignition switch
   I = Off; II = 0; III = On; IV = Start
45 Light switch
46 Hazard flasher switch
47 9 pin plug to turn indicator switch
48 Turn indicator switch
49 6 pin plug to dipswitch
50 Signal button
51 Dipswitch
52 Number plate light and fog lamp connection
53 Left-hand door contact
54 Hazard flasher
55 Interior light
56 Heated rear window connection
57 Selector lever light connection
58 12 pin plug to instrument cluster
59 Heated rear window
60 Selector gate light
61 Fuel gauge transmitter
62 Right-hand door contact
63 Right-hand rear light
   A = Reversing light
   B = Brake light
   C = Flasher
   D = Tail light
64 Number plate light
65 Earth (ground)
66 Left-hand rear light
   A = Reversing light
   B = Brake light
   C = Flasher
   D = Tail light
67 Revolution counter connection
68 Choke
69 Handbrake switch
70 Solder point

Colour code

BL = blue
BR = brown
GE = yellow
GN = green
GR = grey
RT = red
SW = black
VI = violet
WS = white

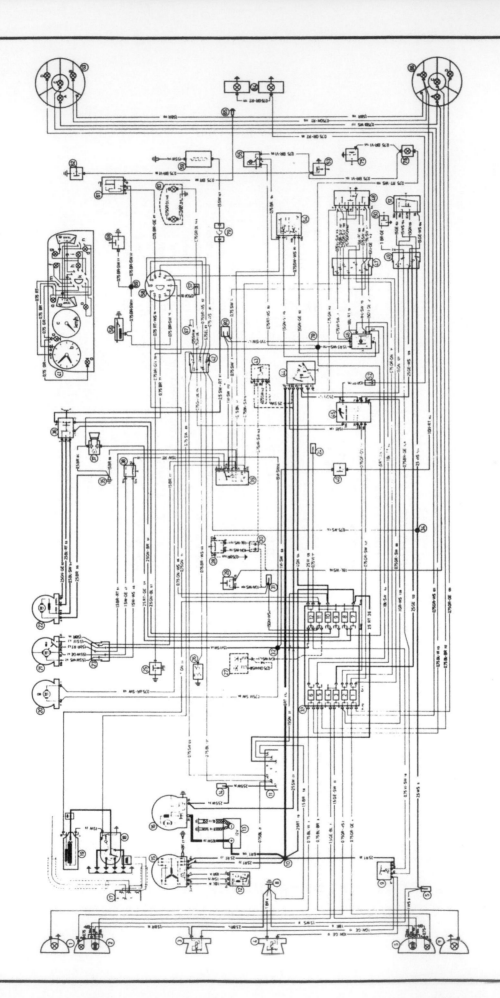

Wiring diagram - 2000 Touring (typical)

Wiring diagram BMW Touring, 2000 Touring

1 Front right-hand flasher
2 RH headlight with parking light
3 Right-hand fanfare
4 Left-hand fanfare
5 Fog lamp relay connection
6 LH headlight with parking light
7 Front left-hand flasher
8 Earth (ground)
9 Fanfare relay
10 Solder point
11 Test equipment connection
12 Regulator
13 Battery
14 Electrical system connection
15 Generator
16 Starter
17 Test equipment connection with lead and pick up
18 Distributor
19 Ignition coil
20 Windshield washer pump
21 Windshield wiper motor
22 Blower motor
23 5 pin plug terminal to wiper motor
24 Radio connection
25 Oil pressure switch
26 Remote rearing therm. trans.
27 Auto. starter carburettor (only with automatic transmission)
28 Reversing light switch with starter lock (only with automatic transmission)
29 Reversing light switch
30 2 pin plug terminal (only with automatic transmission)
31 Fuel pump connection
32 Solder point
33 Fuse box
34 Solder point
35 Earth (ground)
36 Blower switch
37 Cigar lighter

38 Wiper speed switch
39 Wiper-washer transmitter
40 Brake light switch
41 Instrument set
  a) Illuminated scale
  b) Clock
  c) Speedometer
  d) Coolant temperature gauge
  e) Fuel gauge
  f) Charging lamp (red)
  g) Oil pressure control (orange)
  h) Headlight beam telltale (blue)
  i) Flasher control (green)
  k) 12 pin plug
  m) 3 pin plug (clock)
  n) 3 pin plug (rev. counter)
  o) Central control lamp (choke, handbrake, petrol)
42 5 pin plug terminal
43 Starter relay (only with automatic transmission)
44 Ignition switch
  I = Off; II = 0; III = On; IV = Start
45 Light switch
46 Hazard flasher switch
47 9 pin plug to turn indicator switch
48 Turn indicator switch
49 6 pin plug to dipswitch
50 Signal button
51 Dipswitch
52 Number plate light and fog lamp connection
53 Left-hand door contact
54 Hazard flasher
55 Interior light
56 Choke
57 Selector lever connection
58 12 pin plug to instrument cluster
59 Heated rear window

60 Selector gate light
61 Fuel gauge transmitter
62 Right-hand door contact
63 Right-hand rear light
  A = Reversing light
  B = Brake light
  C = Flasher
  D = Tail light
64 Licence plate lights
65 Earth (ground)
66 Left-hand rear light
  A = Reversing light
  B = Brake light
  C = Flasher
  D = Tail light
67 Revolution counter connection
68 Handbrake switch
69 Solder point
70
71
72
73
74
75 Rear door contact
76 LH light in luggage compartment
77 Solder point
78 Solder point
79 Contact plate
80 Rear windshield wiper connection

Colour code
BL = blue
BR = brown
GE = yellow
GN = green
GR = grey
RT = red
SW = black
VI = violet
WS = white

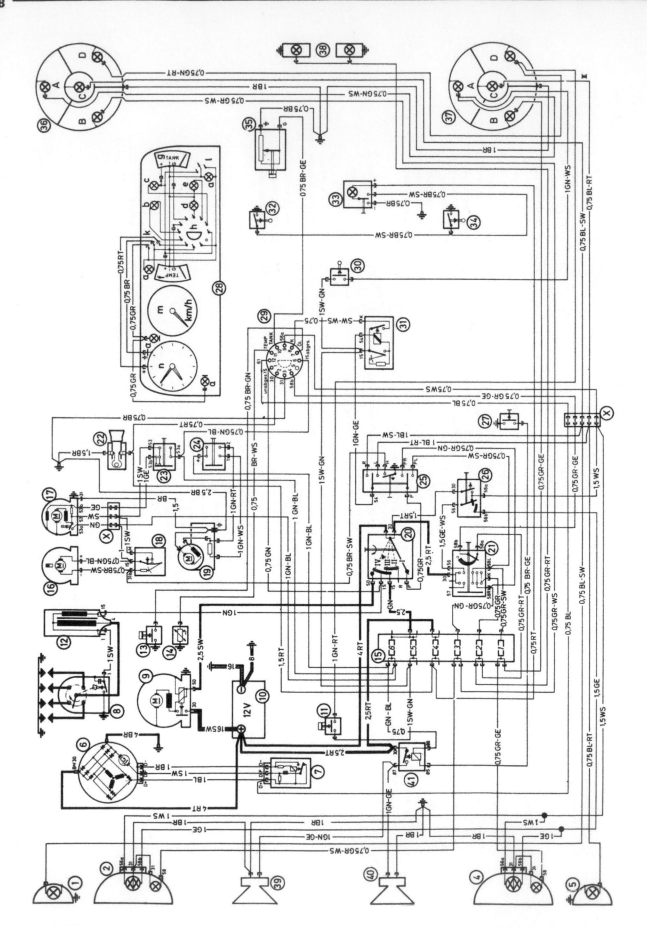

Wiring diagram - 2002 (typical)

Wiring diagram BMW 2002

1 Turn indicator, front RH
2 Headlights, RH with parking light
4 Headlight, LH with parking light
5 Turn indicator, front LH
6 Alternator
7 Regulator
8 Distributor
9 Starter
10 Battery
11 Stop light switch
12 Ignition coil
13 Oil pressure switch
14 Remote thermometer sensor
15 Fuse box
16 Windshield washer pump
17 Windshield wiper motor
18 Delay relay
19 Heater blower
20 Ignition/starter switch
   Switch positions:
   I   Stop
   II Garage
   III Drive
   IV Start

21 Light switch
22 Cigar lighter
23 Windshield wiper switch
24 Heater switch
25 Turn indicator/parking light/windshield washer switch
26 Dipswitch with flasher unit
27 Horn ring
28 Instrument panel
  a) Instrument lighting
  b) Charge telltale (red)
  c) Oil pressure telltale (orange)
  d) Main beam telltale (blue)
  e) Turn indicator telltale (green)
  f) Water temperature gauge
  g) Fuel gauge
  h) 12 pole plug
  k) 3 pole plug (clock)
  l) 3 pole plug (revolution counter)
  m) Speedometer
  n) Clock
29 12 pole round plug for instrument panel (seen from cable side)
30 Reversing light switch
31 Flasher unit
32 Door switch RH
33 Interior light
34 Door switch LH
35 Fuel gauge sensor

36 Rear lights, RH
  A = Reversing light
  B = Rear light
  C = Turn indicator
  D = Stop light
37 Rear lights, LH
  A = Reversing light
  B = Rear light
  C = Turn indicator
38 Number plate light
39 Horn, RH
40 Horn, LH
41 Horn relay
X Flat pin connector

Colour code
BL = blue
BR = brown
GE = yellow
GN = green
GR = grey
RT = red
SW = black
VI = violet
WS = white

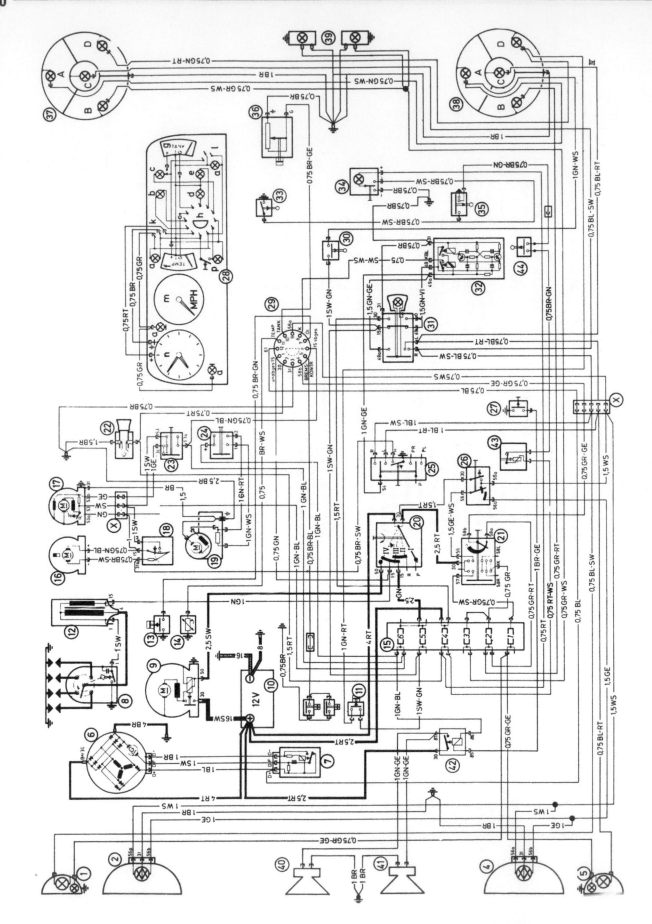

Wiring diagram - 2002 - North America (typical)

Wiring diagram BMW 2002 - North America

1 Turn indicator, front RH with parking light
2 Headlight, RH
4 Headlight, LH
5 Turn indicator, LH with parking light
6 Alternator
7 Regulator
8 Distributor
9 Starter
10 Battery
11 Stop light switch
12 Ignition coil
13 Oil pressure switch
14 Remote thermometer sensor
15 Fuse box
16 Windshield washer pump
17 Windshield wiper motor
18 Delay relay
19 Heater blower
20 Ignition/starter switch - switch positions:
    I   Stop
    II   Garage
    III Drive
    IV Start

21 Light switch
22 Cigar lighter
23 Windshield wiper switch
24 Heater switch
25 Turn indicator/windshield washer switch
26 Dipswitch with flasher unit
27 Horn ring
28 Instrument panel
  a) Instrument lighting
  b) Charge telltale (red)
  c) Oil pressure telltale (orange)
  d) Main beam telltale (blue)
  e) Turn indicator telltale (green)
  f) Water temperature gauge
  g) Fuel gauge
  h) 12 pole plug
  k) 3 pole plug for clock
  l) 3 pole plug for revolution counter
  m) Speedometer
  n) Clock
  p) Brake telltale with test switch
29 12 pole round plug for instrument panel
  (seen from cable side)

30 Reversing light switch
31 Switch for warning flasher unit
32 Warning flasher sensor
33 Door switch, RH
34 Interior light
35 Door switch, LH
36 Fuel gauge sensor
37 Rear lights, RH
  A = Reversing light
  B = Rear light
  C = Turn indicator
  D = Stop light
38 Rear lights, LH
  A = Reversing light
  B = Rear light
  C = Turn indicator
  D = Stop light
39 Number plate light
40 Horn, RH
41 Horn, LH
42 Horn relay
43 Warning buzzer
44 Warning buzzer contact
X Flat pin connector

Colour code

BL = blue
BR = brown
GE = yellow
GN = green
GR = grey
RT = red
SW = black
VI = violet
WS = white

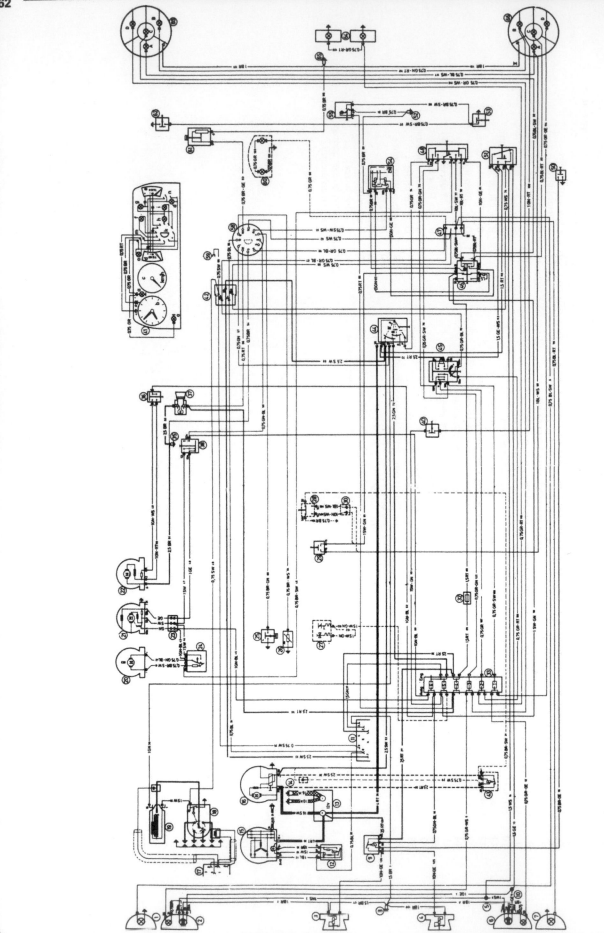

Wiring diagram - 2002 - Automatic transmission (typical)

Wiring diagram BMW 2002 - Automatic

1 Turn indicator, front RH
2 Headlight, RH with parking light
3 Horn, RH
4 Horn, LH
5 Soldering point (56a)
6 Headlight, LH with parking light
7 Turn indicator, front LH
8 Earth
9 Horn relay
10 Soldering point (56b)
11 Connection for diagnosis instrument
12 Regulator
13 Battery
14 Plug connection for starter relay
15 Generator
16 Starter
17 Connection for diagnosis instrument with line and sensor
18 Distributor
19 Ignition coil
20 Windshield washer pump
21 Windshield wiper motor
22 Blower motor
23 3 pole plug connection for wiper motor
24 Delay relay
25 Oil pressure switch
26 Remote thermometer sensor

27 Auto choke carburettor (only with automatic transmission)
28 Reversing light switch with starter lock (only on automatic transmission)
29 Reversing light switch
30 2 pole plug connection (only with automatic transmission)
32 Flying fuse
33 Fuse box
35 Earth
36 Blower switch
37 Cigar lighter
38 Wiper switch
40 Stop light switch
41 Instrument panel
   a) Instrument lighting
   b) Clock
   c) Speedometer
   d) Cooling water temperature gauge
   e) Fuel gauge
   f) Charge telltale (red)
   g) Oil pressure telltale (orange)
   h) Main beam telltale (blue)
   l) Turn indicator telltale (green)
   k) 12 pole plug
   m) 3 pole plug (clock)
   n) 3 pole plug (revolution counter)
42 5 pole plug
43 Starter relay (only with automatic transmission)

44 Ignition/starter switch - switch positions:
   I   Stop
   II 0
   III Drive
   IV Start
45 Light switch
46 Warning flasher switch
47 4 pole plug
48 Turn indicator switch
50 Horn button
51 Dipswitch
52 Earth
53 Door switch, LH
54 Warning flasher sensor
55 Interior light
58 12 pole connector for instrument panel
59 Connector for revolution counter
60 Gate lighting
61 Fuel gauge sensor
62 Door switch, RH
63 Rear lights, RH
   A = Reversing light
   B = Stop light
   C = Turn indicator
   D = Rear light
64 Number plate light
65 Earth
66 Rear lights, LH
   A = Reversing light
   B = Stop light
   C = Turn indicator
   D = Rear light

Colour code
BL = blue
BR = brown
GE = yellow
GN = green
GR = grey
RT = red
SW = black
VI = violet
WS = white

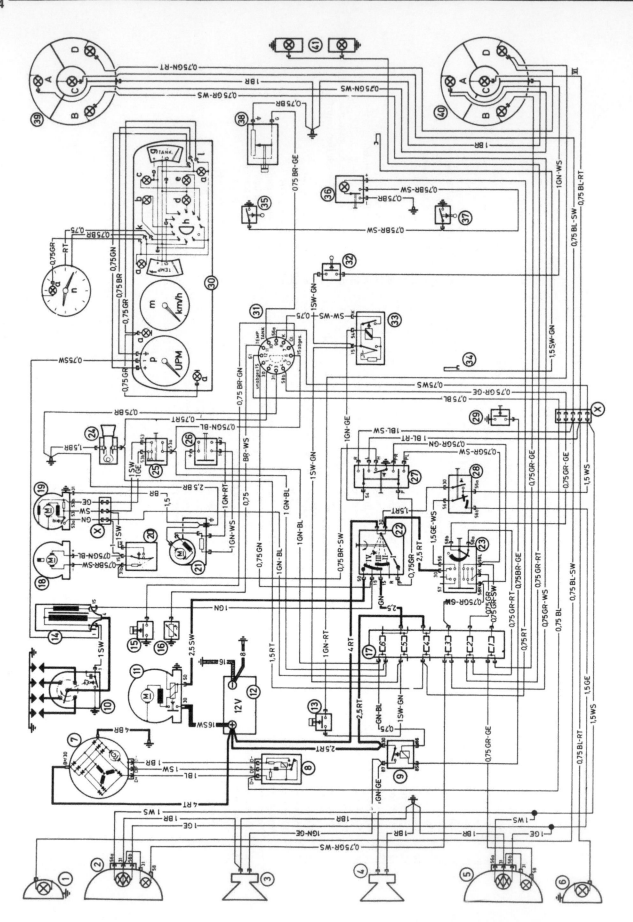

Wiring diagram - 2002 T1 (typical)

Wiring diagram BMW 2002 TI

1 Turn indicator, front RH
2 Headlight, RH with parking light
3 Horn RH
4 Horn LH
5 Headlight, LH with parking light
6 Turn indicator, front LH
7 Alternator
8 Regulator
9 Horn relay
10 Distributor
11 Starter
12 Battery
13 Stop light switch
14 Ignition coil
15 Oil pressure switch
16 Remote thermometer sensor
17 Fuse box
18 Windshield washer pump
19 Windshield wiper motor
20 Delay relay
21 Heater blower

22 Ignition/starter switch - switch positions:
  I   Stop
  II Garage
  III Drive
  IV Start
23 Light switch
24 Cigar lighter
25 Windshield wiper switch
26 Blower switch
27 Turn indicator/parking light/windshield washer switch
28 Dipswitch with flasher unit
29 Horn ring
30 Instrument panel
  a) Instrument lighting
  b) Charge telltale (red)
  c) Oil pressure telltale (orange)
  d) Main beam telltale (blue)
  e) Turn indicator telltale (green)
  f) Water temperature gauge
  g) Fuel gauge
  h) 12 pole plug
  k) 3 pole plug for clock
  l) 3 pole plug for revolution counter
  m) Speedometer
  n) Clock

  p) Revolution counter
31 12 pole round plug for instrument panel (seen from cable side)
32 Reversing light switch
33 Flasher unit
34 Cable for heated backlight (special equipment)
35 Door switch, RH
36 Interior light
37 Door switch, LH
38 Fuel gauge sensor
39 Rear lights, RH
  A = Reversing light
  B = Rear light
  C = Turn indicator
  D = Stop light
40 Rear lights, RH
  A = Reversing light
  B = Rear light
  C = Turn indicator
  D = Stop light
41 Number plate light
X Flat pin connector

Colour code
BL = blue
BR = brown
GE = yellow
GN = green
GR = grey
RT = red
SW = black
VI = violet
WS = white

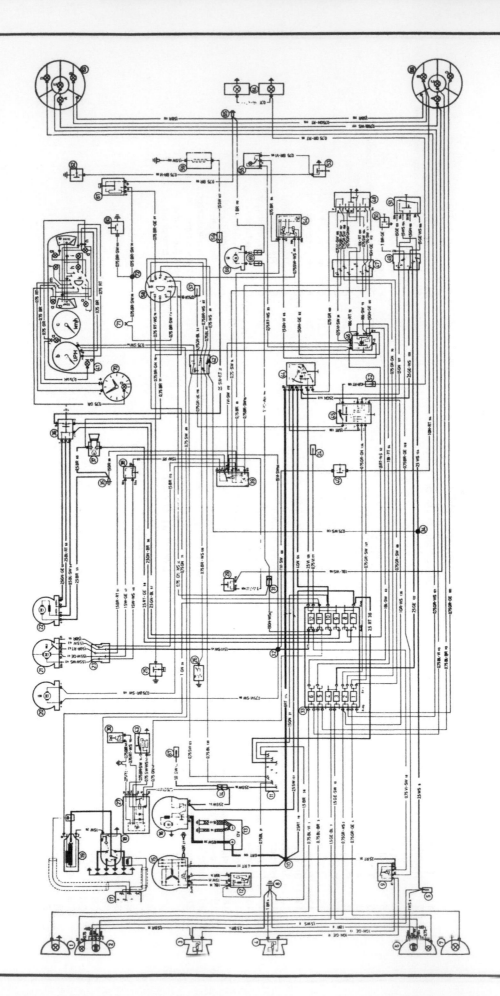

Wiring diagram - 2002 TII (typical)

Wiring diagram BMW 2002 TII

1 Front right-hand flasher
2 RH headlight with parking light
3 Right-hand fanfare
4 Left-hand fanfare
5 Fog lamp relay connection
6 LH headlight with parking light
7 Front left-hand flasher
8 Earth (ground)
9 Fanfare relay
10 Solder point
11 Test equipment connection
12 Regulator
13 Battery
14 Connection for electrical system
15 Generator
16 Starter
17 Test equipment connection with lead and pick up
18 Distributor
19 Ignition coil
20 Windshield washer pump
21 Windshield wiper motor
22 Blower motor
23 5 pin plug to wiper motor
24 Connection for radio
25 Oil pressure switch
26 Remote thermometer transmitter
27 Time switch
29 Reversing light switch
30 Starting valve
31 Fuel pump connection
32 Solder point
33 Fuse box
34 Solder point
35 Earth (ground)
36 Blower switch
37 Cigar lighter
38 Wiper speed switch
39 Wiper-washer transmitter
40 Brake light switch

41 Instrument cluster
  a) Illuminated scale
  b) Revolution counter
  c) Speedometer
  d) Coolant temperature gauge
  e) Fuel gauge
  f) Charging lamp (red)
  g) Oil pressure switch (orange)
  h) High beam telltale (blue)
  i) Flasher control (green)
  k) 12 pin plug
  m) 3 pin plug (clock)
  n) 3 pin plug (revolution counter)
  o) Central control lamp (choke, handbrake, petrol)
42 5 pin plug terminal
43 Temperature time switch
44 Ignition switch
  I = Off; II = 0; III = On; IV = Start
45 Light switch
46 Hazard flasher switch
47 9 pin plug to turn indicator switch
48 Turn indicator switch
49 6 pin plug to dipswitch
50 Signal button
51 Dipswitch
52 Number plate light and fog lamp connection
53 Left-hand door contact
54 Hazard flasher
55 Interior light
56 Connection for heater rear window
57 Selector lever light connection
58 12 pin plug to instrument cluster
59 Heated rear window (SA)
60 Handbrake switch
61 Fuel gauge transmitter
62 Right-hand door contact

63 Right-hand rear light
  A = Reversing light
  B = Brake light
  C = Flasher
  D = Tail light
64 Number plate light
65 Earth (ground)
66 Left-hand rear light
  A = Reversing light
  B = Brake light
  C = Flasher
  D = Tail light
67 HT ignition system connection
68 Fuel pump
69 Plug connection to fuel pump
70 Clock
71 Choke connection
72 Solder point

Colour code
BL = blue
BR = brown
GE = yellow
GN = green
GR = grey
RT = red
SW = black
VI = violet
WS = white

168

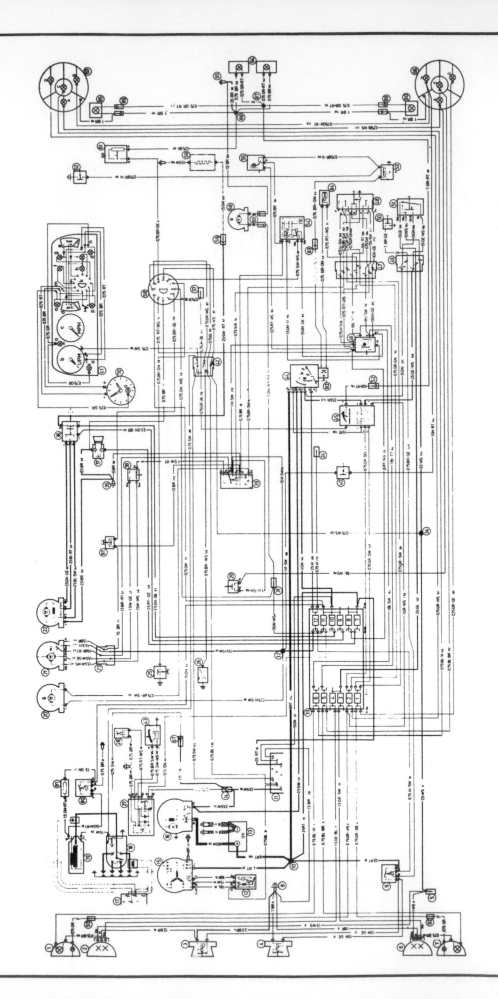

Wiring diagram BMW 2002 TII - North America

1 Front RH flasher with parking light
2 Right-hand headlight
3 Right-hand fanfare
4 Left-hand fanfare
5 Fog lamp relay connection
6 Left-hand headlight
7 Front LH flasher with parking light
8 Earth (ground)
9 Fanfare relay
10 Solder point
11 Test equipment connection
12 Regulator
13 Battery
14 Connection for electrical equipment
15 Generator
16 Starter
17 Test equipment connection with lead and pick up
18 Distributor
19 Ignition coil
20 Windshield washer pump
21 Windshield wiper motor
22 Blower motor
23 5 pin plug terminal to wiper motor
24 Radio connection
25 Oil pressure switch
26 Remote reading therm. transmitter
27 Time switch
28 Starting valve
29 Reversing light switch
30
31 Fuel pump connection
32 Solder point
33 Fuse box
34 Solder point
35 Earth (ground)
36 Blower switch
37 Cigar lighter
38 Wiper speed switch
39 Wiper washer transmitter

40 Brake light switch
41 Instrument cluster
   a) Illuminated scale
   b) Revolution counter
   c) Speedometer
   d) Coolant temperature gauge
   e) Fuel gauge
   f) Charging lamp (red)
   g) Oil pressure telltale (orange)
   h) High beam telltale (blue)
   i) Flasher telltale (green)
   k) 12 pin plug
   m) 3 pin plug (clock)
   n) 3 pin plug (rev. counter)
   p) Brake fluid control (red)
42 5 pin plug terminal
43 Temperature time switch
44 Ignition switch
   I = Off; II = 0; III = On; IV = Start
45 Light switch
46 Hazard flasher switch
47 9 pin plug terminal to turn indicator switch
48 Turn indicator switch
49 6 pin plug terminal to dipswitch
50 Signal button
51 Dipswitch
52 Number plate light and fog lamp connection
53 Door double contactor
54 Hazard flasher
55 Interior light
56 Heated rear window connection
57 Selector lever light connection
58 12 pin plug terminal to instrument cluster
59 Heated rear window
60
61 Fuel gauge transmitter
62 Door contact

63 Right hand rear light
   A = Reversing light
   B = Brake light
   C = Flasher
   D = Tail light
64 Number plate lights
65 Earth (ground)
66 Left-hand rear light
   A = Reversing light
   B = Brake light
   C = Flasher
   D = Tail light
67 HT ignition system connection
68 Fuel pump
69 Fuel pump plug connection
70 Clock
71
72
73
74
75
76
77
78
79
80 HT ignition system relay
81 Series resistance
82 Buzzer contact
83 Buzzer contact connection
84 Buzzer
85 RH side marker light
86 RH side marker light connection
87 Solder point 58 K
88 Solder point 31
89 LH side marker light connection
90 Left-hand side marker light
91 Brake fluid level switch
92 Cable connector
93 Cable connector

Colour code

BL = blue
BR = brown
GE = yellow
GN = green
GR = grey
RT = red
SW = black
VI = violet
WS = white

Chapter 11 Suspension and steering

Contents

Specifications

Front suspension

Type	Independent. MacPherson strut, coil spring and stabiliser bar on certain models

	1500, 1600	1502, 1602	2000 Touring	2002
Coil springs				
Free length	12.11 in (307.5 mm)	early: 12.78 in (324.8 mm) late: 13.12 in (333.2 mm)	13.58 in (345.0 mm)	early: 13.12 in (333.2 mm) late: 12.78 in (324.8 mm)
Wire diameter	0.495 in (12.5 mm)	0.465 in (11.8 mm)	0.473 in (12.0 mm)	0.465 in (11.8 mm)
Coil diameter (external)	5.02 in (127.5 mm)	4.99 in (126.8 mm)	4.99 in (126.8 mm)	4.99 in (126.8 mm)
Diameter of stabiliser bar	0.67 in (17.0 mm)	0.59 in (15.0 mm)	0.67 in (17.0 mm)	0.59 in (15.0 mm)
Hub bearing endfloat (all models)	0.0008 to 0.002 in (0.02 to 0.06 mm)			
Track	1500, 1600	1502, 1602	2000 Touring	2002
... ...	52.0 in (1320.0 mm)	52.4 in (1330.0 mm)	52.4 in (1330.0 mm)	52.4 in (1330.0 mm)
Camber (positive) *	0 to 1º	0 to 1º	0 to 1º	0 to 1º
Kingpin inclination *	8º 30'	8º 30'	8º 30'	8º 30'
Castor *	3º 30' to 4º 30'	3º 30' to 4º 30'	3º 30' to 4º 30'	3º 30' to 4º 30'
Toe-in	0 to 0.079 in (0 to 2 mm)	0.039 to 0.81 in (1 to 2.5 mm)	0.39 to 0.81 in (1 to 2.5 mm)	0.39 to 0.81 in (1 to 2.5 mm)

* Not adjustable

Rear suspension

Type ... ...   Independent, trailing arms pivoted on sub-frame, coil springs and double acting telescopic shock absorbers. Stabiliser bar on certain models

Coil springs	**1500, 1600**	**1502, 1602**	**2000 Touring**	**2002**
Free-length	11.38 in (289.0 mm)	13.17 in (334.6 mm)	13.57 in (344.8 mm)	14.25 in (362.0 mm)
Wire diameter	0.56 in (14.2 mm)	0.49 in (12.3 mm)	0.504 in (12.8 mm)	0.46 in (11.8 mm)
Coil diameter (external)	4.7 in (119.2 mm)	5.01 in (127.3 mm)	5.03 in (127.8 mm)	4.99 in (126.8 mm)

Diameter of stabiliser bar	0.67 in (17.0 mm)	0.63 in (16.1 mm)	0.67 in (17.0 mm)	0.63 in (26.8 mm)

Hub bearing endfloat (set during assembly) ... ...   0.002 to 0.004 in (0.05 to 0.1 mm)

Track	53.8 in (1366.0 mm)	52.4 in (1330.0 mm)	54.2 in (1376.0 mm)	52.4 in (1330.0 mm) TII model: 53.0 in (1348.0 mm)
Camber *	2° neg	1° 30' to 2° 30' neg	2° neg	1° 30' to 2° 30' neg
Toe-in *	0 to 0.078 in (0 to 2 mm)	0 to 0.12 in (0 to 3 mm)	0 to 0.078 in (0 to 3 mm)	0 to 0.078 in (0 to 3 mm)

* Not adjustable

Steering

Type ... ...   Worm and roller with idler, rod and two outer trackrods

Ratio ... ...   15.5 : 1

Number of turns of steering wheel lock-to-lock 3½

Maximum wheel lock
Inner ... ...   42°
Outer ... ...   34°

Turning circle ... ...   34.1 ft (10.4 m)

Wheels and tyres	**1500, 1600**	**1502, 1602**	**2000 Touring**	**2002**
Roadwheels				
Type (all models)		Steel disc		
Size	4½J X 14	4½J X 13	5J X 14	5J X 13
Tyres				
Size (crossply)	6.00 - 14	6.00 - 13	6.45 - 14	6.45 - 14
Pressures (crossply):				
Front	24 psi (1.7 atm)	24 psi (1.7 atm)	(24 psi (1.7 atm)	—
Rear	24 psi (1.7 atm)	24 psi (1.7 atm)	24 psi (1.7 atm)	—
Size (radial)	165 SR 14	165 SR 13	165 SR 14	165 SR 13 or 165 HR 13
Pressures (radial):				
Front	26 psi (1.8 atm)	26 psi (1.8 atm)	26 psi (1.8 atm)	26 psi (1.8 atm)
Rear	26 psi (1.8 atm)	26 psi (1.8 atm)	26 psi (1.8 atm)	26 psi (1.8 atm)

Torque wrench settings	**lb/ft**	**Nm**
Front suspension		
Crossmember bolts	52	72
Suspension strut top mounting nuts	20	28
Suspension strut piston rod nut	55	76
Suspension strut threaded ring	85	118
Track control arm pivot bolt	110	152
Radius rod nuts	50	69
Stabiliser clamp bolts	18	25
Stabiliser end nuts	18	25

Caliper to stub axle carrier	...	...	...	...	...	...	...	70	97
Steering arm to track control arm castellated nut	...	...	...	...	...		30	41	
Steering arm bolts to stub axle carrier	...	...	...	...	...		25	35	
Balljoint nuts	...	...	...	...	...	...	...	32	44

Rear suspension

Shock absorber lower mounting	...	...	...	...	...	...	38	53	
Trailing arm pivot bolts	...	...	...	...	...	...	...	55	76
Subframe mounting bolts to body	...	...	...	...	...	...	110	152	
Subframe tie-bar bolts to body	...	...	...	...	...	...	32	44	
Final drive housing to subframe mounting bolts	...	...	...	...	56	77			
Final drive to rear bracket bolts	...	...	...	...	...		38	53	
Rear bracket to body bolts	...	...	...	...	...	...	38	53	
Driveshaft nut	...	...	...	...	...	...	...	217	294

Steering

Steering wheel nut:

12 mm	...	...	...	...	...	...	...	...	40	55
14 mm	...	...	...	...	...	...	...	...	62	86
Flexible coupling disc bolts	...	...	...	...	...	...	18	25	
Flexible coupling pinch bolts	...	...	...	...	...	...	20	28	
Drop arm nut	...	...	...	...	...	...	...	90	125	
Idler castellated nut	...	...	...	...	...	...	...	60	83	
Steering box end cover bolts	...	...	...	...	...	...	14	19	
Steering box mounting bolts	...	...	...	...	...	...	34	47	
Trackrod clamp bolts	...	...	...	...	...	...	...	10	14	
Roadwheel nuts	...	...	...	...	...	...	...	65	90	

1 General description

The front suspension is of independent type, incorporating MacPherson struts. A stabiliser bar is fitted to some models.

The rear suspension is of independent type incorporating coil springs telescopic shock absorbers and rubber bushed semi-trailing arms.

The steering gear is of worm and roller type with a three piece trackrod.

Radial ply tyres are fitted to all later models although crossply were fitted as standard until 1971.

2 Maintenance and inspection

1 Regularly inspect the condition of all flexible gaiters, balljoint boots and suspension bushes for wear or deterioration.

2 Check the security of all steering and suspension nuts and bolts at frequent intervals, particularly the nuts of the front suspension strut upper mountings.

3 At the intervals specified in 'Routine Maintenance', clean, repack with lubricant and adjust the wheel bearings and check the front wheel alignment. Top-up the steering box with the specified grade of lubricant.

3 Front hub bearings - adjustment

1 Jack-up the front of the car and remove the roadwheels.

2 Tap or twist off the dust cap and extract the split pin.

3 Withdraw the disc pads (Chapter 9).

4 Rotate the hub in its normal (forward) direction of travel and tighten the castellated nut to a torque of 20 lb/ft (28 Nm) to settle the bearings.

5 Release the nut and tighten it with the fingers only. Bearing endfloat should have been eliminated or certainly should not exceed 0.02 mm which is an almost imperceptible amount. Do not confuse endfloat with rocking of the taper roller bearings which will be evident if the bearings are worn.

6 If the adjustment has been carried out correctly, the thrust washer behind the castellated nut should be just free enough to be rotated slightly in either direction.

7 Install a new split pin and bend the ends well over. Refit the dust cap and the disc pads.

8 Repeat the operations on the opposite hub and then refit the roadwheels and lower the jack.

4 Front hub bearings - removal, refitting and lubrication

1 Jack-up the front of the car, remove the roadwheel. Disconnect the fluid line bracket from the suspension strut, unbolt the caliper and support it to one side, there is no need to disconnect the hydraulic lines.

2 Tap or twist off the dust cap, remove the split pin and unscrew and remove the castellated nut.

3 Extract the thrust washer and then pull the hub assembly carefully from the stub axle. Catch the outer tape bearing.

4 Using a suitable drift drive out the oil seal and extract the inner taper roller bearing.

5 If new bearings are to be installed, drive out the bearing tracks using a bearing puller or drift.

6 Clean the interior of the hub and dust cap and then install the new bearing tracks, the inner taper roller bearing and tap in a new oil seal.

7 Pack wheel bearing grease into the inside of the hub so that it fills the deeper recess but is no thicker than the bearing tracks. Press some grease into the bearing rollers and then reassemble by reversing the dismantling process.

8 Adjust the bearings, as described in Section 3 and refit the dust cap. Do not fill the cap more than 1/3rd full with grease.

5 Front suspension assembly - removal and installation

1 Jack-up the front of the car and support it under the side bodyframe members just to the rear of the front crossmember.

2 Remove the roadwheels.

3 Disconnect the support bracket halfway up, and to the rear, of each suspension strut, unbolt the caliper units and pull them to one side and tie them up out of the way. There is no need to disconnect the hydraulic fluid lines.

4 Loosen both the pinch bolts on the steering shaft flexible coupling.

5 Take the weight of the engine on a suitable hoist and remove the engine front mountings.

6 Unscrew and remove the upper mounting nuts from the suspension struts.

7 Support the front crossmember on a jack, preferably trolley type and then unbolt the crossmember from the bodyframe and lower and remove the complete suspension assembly. Note that the steering flexible coupling should be eased from the shaft splines as the suspension is lowered.

8 Installation is a reversal of removal but tighten all nuts and bolts to the specified torque and make sure that the flexible coupling is connected when the steering wheel and the front hubs are in the straight-ahead position. Adjust the engine mounting stop to give a clearance of 0.12 in. (3.0 mm) as shown.

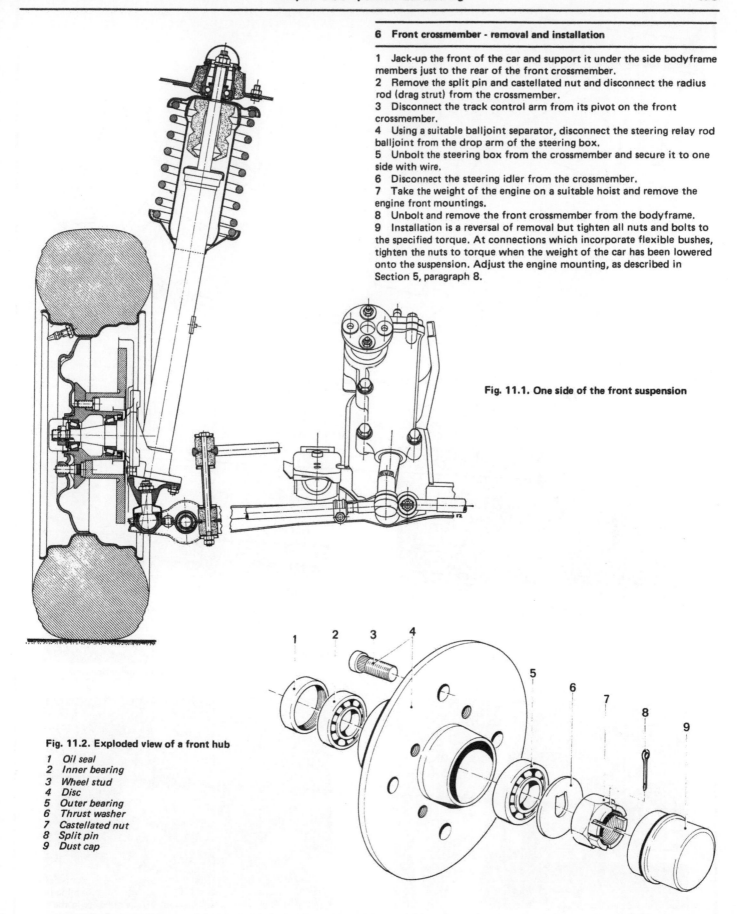

6 Front crossmember - removal and installation

1 Jack-up the front of the car and support it under the side bodyframe members just to the rear of the front crossmember.
2 Remove the split pin and castellated nut and disconnect the radius rod (drag strut) from the crossmember.
3 Disconnect the track control arm from its pivot on the front crossmember.
4 Using a suitable balljoint separator, disconnect the steering relay rod balljoint from the drop arm of the steering box.
5 Unbolt the steering box from the crossmember and secure it to one side with wire.
6 Disconnect the steering idler from the crossmember.
7 Take the weight of the engine on a suitable hoist and remove the engine front mountings.
8 Unbolt and remove the front crossmember from the bodyframe.
9 Installation is a reversal of removal but tighten all nuts and bolts to the specified torque. At connections which incorporate flexible bushes, tighten the nuts to torque when the weight of the car has been lowered onto the suspension. Adjust the engine mounting, as described in Section 5, paragraph 8.

Fig. 11.1. One side of the front suspension

Fig. 11.2. Exploded view of a front hub

1   Oil seal
2   Inner bearing
3   Wheel stud
4   Disc
5   Outer bearing
6   Thrust washer
7   Castellated nut
8   Split pin
9   Dust cap

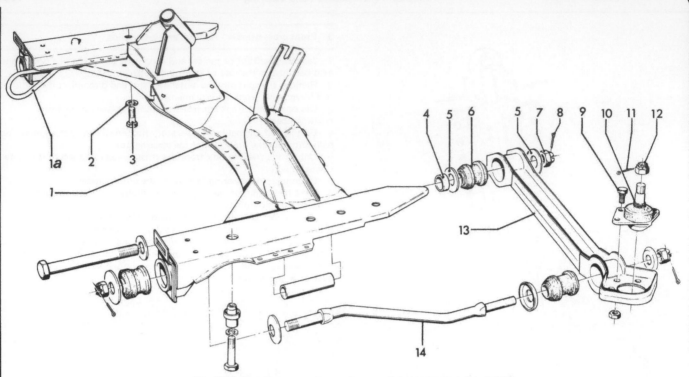

Fig. 11.3. Front crossmember and suspension components (one side)

1	Crossmember	4 Spacer	8 Split pin	12 Nut
1a	Towing eye	5 Washer	9 Bolt	13 Track control arm
2	Washer	6 Flexible bush	10 Balljoint	14 Radius rod (drag
3	Bolt	7 Castellated nut	11 Split pin	strut)

Fig. 11.4. Steering shaft flexible coupling

1 and 2   Pinch bolts

Fig. 11.5. Suspension strut upper mounting nuts

Fig. 11.6. Front crossmember mounting bolts (one side)

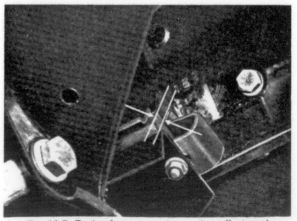

Fig. 11.7. Engine front mounting setting adjustment

Fig. 11.8. Radius rod attachment to crossmember

Fig. 11.9. Track control arm attachment to crossmember

7 Track control arm - removal and refitting

1 Jack-up the front of the car and remove the roadwheel.
2 Remove the hub assembly and disc shield.
3 Disconnect the stabiliser bar from the track control arm.
4 Disconnect the track control arm from the front crossmember pivot.
5 Using a suitable balljoint separator, disconnect the trackrod-end from the steering arm of the stub axle carrier.
6 Disconnect the radius rod from the track control arm.
7 Cut the wire seal and unbolt and separate the track control arm/steering arm assembly from the base of the suspension strut/stub axle carrier.
8 After unscrewing and removing the castellated nut, the swivel joint can be pressed from the steering arm. The joint can be removed from the track control arm by drilling out the securing rivets. A new joint is supplied complete with nuts and bolts, make sure that the nuts are located on the side opposite to the taper pin.
9 If the pivot bushes are worn on the track control arm, press them out and install new ones. A long bolt and washers and distance pieces can be used to remove and install the bushes. A little glycerine or hydraulic fluid will assist installation of the new bushes.
10 Refitting is a reversal of removal but make sure that the washers which are located where the radius rod connects with the track control arm are so fitted that they engage with the curved surfaces of the track control arm. Make sure also that the spacer is correctly located between the track control arm and the crossmember when the arm is fitted to its pivot. Tighten this nut to torque when the car has been lowered onto the suspension.

8 Radius rod (drag strut) - removal and refitting

1 Jack-up the front of the car and disconnect the track control arm from its pivot on the front crossmember.
2 Disconnect the radius rod from the track control rod and then unbolt it from the bracket on the front crossmember.
3 The flexible mounting bush in the bracket can be renewed by using a bolt, washers and suitable distance pieces but the stabiliser bar clamp will have to be unbolted first.
4 Refitting is a reversal of removal but tighten the radius rod nuts to specified torque after the weight of the car has been lowered onto the suspension.

9 Front stabiliser bar - removal and refitting

1 Disconnect the stabiliser bar from the track control arm by removing the nut and locknut.
2 Unbolt the flexible clamps and remove the bar.
3 Refitting is a reversal of removal but tighten all nuts and bolts to specified torque when the weight of the car is on the suspension.

10 Front suspension strut - removal and installation

1 Jack-up the front of the car, remove the hub and disc shield.
2 Disconnect the trackrod end from the steering arm on the base of the suspension strut using a suitable balljoint separator.
3 Disconnect the track control arm from its pivot on the front crossmember.
4 Cut the locking wire and disconnect the track control arm swivel joint from the base of the suspension strut (three bolts).
5 Remove the strut upper mounting nuts and withdraw the strut downwards and out from under the front wing.
6 Installation is a reversal of removal.

11 Front suspension strut - overhaul

1 With the strut removed from the car, compress the coil spring using screw type compressors.
2 Prise off the cap from the top mounting and hold the piston rod quite still with a spanner and unscrew the self-locking nut. Withdraw the top mounting.
3 Withdraw the coil spring (still compressed) together with its upper mounting plate.
4 If the coil spring is to be renewed, release the compressors and fit them to the new spring and compress it ready for refitting. Always renew both front coil springs at the same time with ones of similar rating and colour identification.
5 Renewal of the suspension hydraulic strut can be carried out in one of two ways, either renew the strut complete with a new or reconditioned unit or renew the internal components with a repair kit or cartridge.
6 To dismantle the strut, unscrew the cap and unscrew the threaded ring from the top of the strut tube. A special tool is needed to unscrew this ring but if one cannot be borrowed or made up, careful use of a cold chisel and hammer will provide a good substitute.
7 Prise up the collar now exposed and lift out the 'O' ring seal.
8 Extract the internal piston assembly and thoroughly clean out the interior of the strut tube.
9 Install the repair kit piston assembly and inject the 285 cc of special oil supplied into the inner tube.
10 Refit the new 'O' ring, collar and screw on the threaded ring by hand only.
11 Pull the piston rod fully out to its stop and tighten the threaded ring to 85 lb/ft (118 Nm).
12 Refit the remaining components including the coil spring in its compressed state and making sure that the ends of the spring are correctly located in the recesses in the spring caps. Note that the conical end of the auxiliary spring must point downwards and the packing ring above the upper spring cap must have its turned over edges pointing upwards.
13 Holding the end of the piston rod quite still with a spanner, tighten the self-locking nut to a torque of 55 lb/ft (76 Nm).

Fig. 11.10. Stabiliser bar attachment to track control arm

Fig. 11.11. Suspension lower swivel joint securing bolts

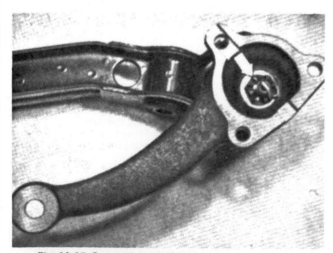

Fig. 11.12. Suspension swivel joint to steering arm nut

Fig. 11.13. Bolting a new suspension balljoint to the track control arm

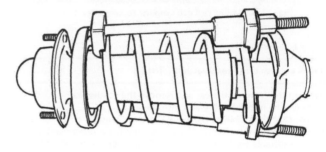

Fig. 11.14. Suspension strut coil spring compressed

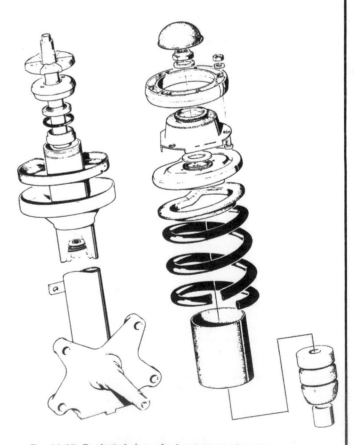

Fig. 11.15. Exploded view of a front suspension strut

12 Rear stub axle - removal and refitting

1   Apply the handbrake and remove the hub cap.
2   Extract the split pin then unscrew the castellated nut. Also loosen the roadwheel nuts.
3   Release the handbrake, jack up the car, and remove the wheel and brake drum.
4   Using a suitable two or three-legged extractor, pull off the driving flange.
5   Disconnect the driveshaft from the flange on the inner face of the hub. Tie the driveshaft up out of the way.
6   Screw on the castellated nuts two or three turns to protect the threads and then drive the splined stub axle out of the hub by striking it with a plastic faced mallet.
7   Refitting is a reversal of removal, but the following should be noted:
   a) If new oil seals have been used, pack the sealing grooves with graphite grease
   b) With the roadwheel fitted, handbrake applied and the car lowered to the ground, tighten the wheel nuts and castellated driveshaft nut to their specified torques

13 Rear hub bearings - dismantling, lubrication and reassembly

1   Remove the stub axle, as described in the preceding Section.
2   Remove the inner and outer oil seals and bearings using either a bearing extractor or driving them out with a suitable drift. Extract the shim(s) and tubular spacer.
3   If new bearings are to be fitted, install the inboard one first then, using a depth gauge, measure the distance (B) between the seat of the outboard bearing and the face of the inboard bearing outer track.
4   Using a vernier gauge, measure the length (A) of the spacer.
5   Subtract B from A and from the result subtract the specified bearing endfloat (see Specifications). The final figure represents the thickness of shims which must be installed to the inner face of the outer bearing.
6   Reassemble the components in the reverse order to removal, having injected 1.2 oz (35 g) of wheel bearing grease into the bearings and hub. Pack the sealing grooves of new oil seals with graphite grease.
7   Finally, with the roadwheel fitted, handbrake applied and the car lowered to the ground, tighten the wheel nuts and castellated driveshaft nut to their specified torques.

14 Rear shock absorber - removal, testing and refitting

1   Jack-up the rear of the car and support it securely.
2   Using a second jack, support the suspension trailing arm otherwise the driveshaft universal joints will be strained when the shock absorbers (which act as check links) are released from their mountings.
3   Working inside the luggage boot, remove the cap and disconnect the shock absorber upper mounting.
4   Disconnect the shock absorber lower mounting and lift the unit from the car.
5   To test the shock absorber, grip its lower mounting eye in the jaws of a vice and with the unit in the vertical position, fully extend and compress it about ten times. If at any time, any lack of resistance is felt or if the shock absorber locks then it must be renewed.
6   Refitting is a reversal of removal but make sure that the upper conical flexible mounting has the narrower diameter towards the shock absorber and tighten the mounting nuts to the specified torque only after the weight of the car has been lowered onto the suspension.

15 Rear stabiliser bar - removal and refitting

1   This is simply a matter of unbolting the clamps and the end fittings and removing the bar from the cap (photos).
2   When refitting, do not tighten the securing bolts to specified torque until the weight of the car is on the suspension.

16 Rear coil spring - removal and refitting

1   Remove the rear shock absorber (Section 13) and the roadwheel. Support the suspension trailing arm with a jack placed under it.
2   Disconnect the stabiliser bar from the trailing arm of the suspension.

3   Disconnect the driveshaft at its outer flange and tie the shaft up out of the way.
4   Lower the jack under the suspension and withdraw the coil spring.
5   If the spring is to be renewed, always fit a spring of similar rating to that of the original and of similar colour coding.
6   Renew the spring insulator rings if the old ones are damaged.
7   Refitting is a reversal of removal but make sure that the ends of the springs locate correctly on their insulators.

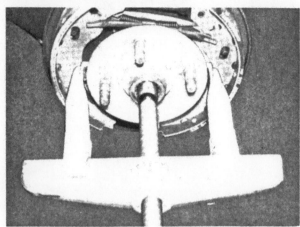

Fig. 11.16. Removing a rear hub driving flange

Fig. 11.17. Driveshaft disconnected from stub axle flange

Fig. 11.18. Removing stub axle from hub

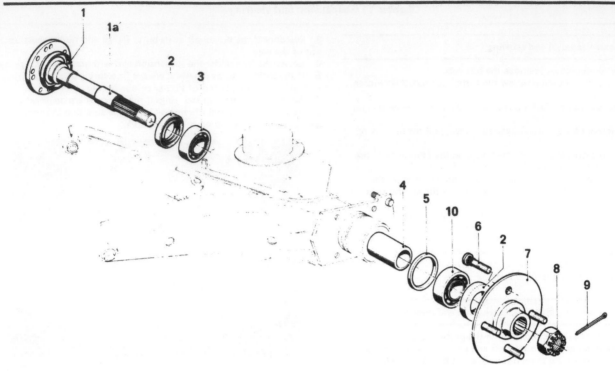

Fig. 11.19. Exploded view of a rear hub

1 Dust deflector	3 Inboard bearing	6 Wheel stud	9 Split pin
1a Stub axle	4 Tubular spacer	7 Driving flange	10 Outboard bearing
2 Oil seal	5 Shim	8 Castellated nut	

Fig. 11.20 Measuring distance between bearings

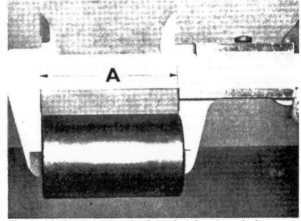

Fig. 11.21 Measuring length of rear hub bearing tubular spacer

Fig. 11.22. Disconnecting rear shock absorber upper mounting

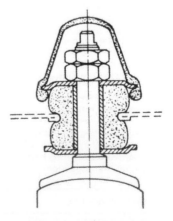

Fig. 11.23. Sectional view of rear shock absorber upper mounting

Fig. 11.24. Correct fitting of rear shock absorber flexible bush

Fig. 11.25. Rear coil spring insulator ring (note locking notch)

15.1a Rear stabilizer bar end attachment

15.1b Rear stabilizer bar clamp and rear suspension sub frame

17 Rear subframe - removal and installation

1   Removal of the subframe upon which the front of the differential is mounted and the rear suspension trailing arms are pivoted, is carried out in the following way.
2   Raise the rear of the car and support it securely on stands placed under the body sideframe members.
3   Slacken the bolts on the differential front flexible mountings, the trailing arm pivot bolts and the subframe mounting bolts.
4   Remove the stabiliser bar clamps from the subframe.
5   With jacks or stands, support the trailing arms under the coil spring seats and under the inner arm. Support the differential unit.
6   Unbolt and remove the two subframe to body tie-bars. Remove the previously released subframe mounting bolts and lower the subframe from the car.
7   If the subframe flexible mounting bushes require renewal, the rear seat will have to be removed and the knurled bearing studs driven out. When fitting the new bushes make sure that the projecting collar faces upwards.
8   Installation is a reversal of removal but tighten all bolts and nuts to their specified torque when the weight of the car is on the suspension.

18 Rear suspension trailing arm - removal and refitting

1   Remove the coil roadspring (Section 16).
2   Disconnect the appropriate cable from the handbrake lever and from the suspension arm clips.
3   Disconnect the stabiliser bar from the suspension trailing arm.
4   Disconnect the flexible brake hose from the rigid brake line.
5   Disconnect the trailing arm from both its pivots on the subframe and withdraw it complete with hub assemlby.
6   The trailing arm bushes can be renewed by extracting and inserting them using a bolt, washers and distance pieces.
7   Refitting is a reversal of removal but tighten bolts to specified torque only when weight of car is on the suspension.
8   Bleed the brakes, as described in Chapter 9.

19 Steering wheel - removal and refitting

1   Set the ignition key in the 'GARAGE' position and turn the steering wheel to the 'straight-ahead' position.
2   Remove the centre cover from the wheel, either by extracting the screws from the rear of the cover or by prising it off according to type.
3   Unscrew and remove the steering wheel securing nut and pull the wheel from the shaft splines. The wheel will normally come off if it is jarred with the palms of the hands behind its rim (photo).
4   Refit the wheel in its original 'straight-ahead' position making sure that the front roadwheels have not been turned. An alignment mark is usually to be found on the end of the steering shaft and the steering wheel hub.

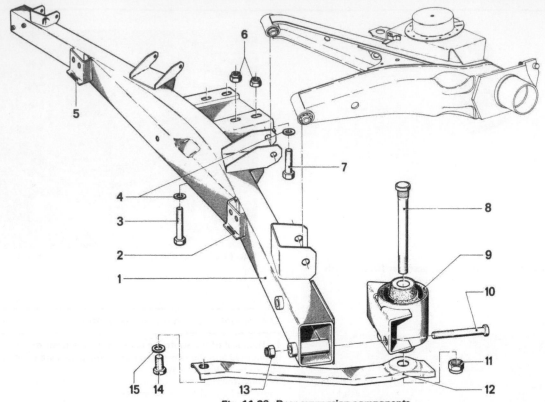

Fig. 11.26. Rear suspension components

1	Subframe	5	Bearing support	9	Mounting	13	Nut
2	Bearing support	6	Nuts	10	Bolt	14	Bolt
3	Bolt	7	Bolt	11	Nut	15	Washer
4	Washer	8	Bolt	12	Tie bar	16	Trailing arm

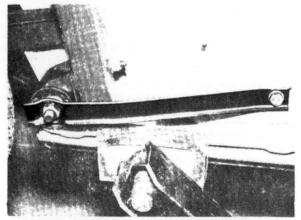

Fig. 11.27. Rear subframe to body tie-bars

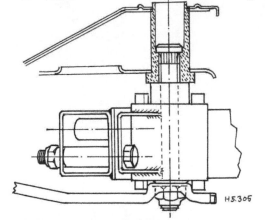

Fig. 11.28. Sectional view of a subframe mounting bush assembly

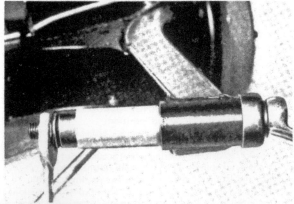

Fig. 11.29. Method of renewing a rear suspension trailing arm flexible bush

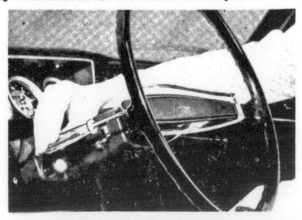

Fig. 11.30. Extracting a steering wheel cover screw

Fig. 11.31. Prising out the steering wheel hub cover

Fig. 11.32. Steering column shroud screws

19.3 Steering wheel securing nut and horn switch leads

Fig. 11.33. Steering column lower shroud showing choke control
with knob removed

20 Steering shaft - removal and refitting

1   Remove the steering wheel, as described in the preceding Section.
2   Unscrew and remove the screws which retain the upper and lower
steering column shrouds together. Remove the upper shroud and after
unscrewing the choke knob and cable locknut, withdraw the lower
shroud.
3   Remove the direction indicator switch from its support bracket.
4   Slacken the upper pinch bolt of the flexible coupling and withdraw
the steering shaft upwards from the column into the car interior.
5   Refitting is a reversal of removal but set the direction indicator
switch so that the gap between the dog and reset cam is 0.012 in (0.3
mm).
6   Before tightening the pinch bolt on the flexible coupling, have an
assistant push (hard) downwards on the steering wheel and lock the
pinch bolt while the shaft is still under load. This will apply the
specified preload.

21 Steering shaft - dismantling and reassembly

1   With the shaft removed, remove the collar to expose the circlip.
Extract the circlip, washer, spring, ring and bearing.
2   Lift the retaining ring from its groove.
3   Renew worn parts and refit them in their correct sequence making
sure that the flange of the ring '4' (Fig. 11.35), is towards the top of
the shaft.

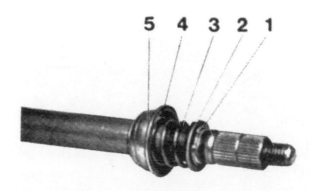

Fig. 11.34. Upper end of steering shaft

1 Circlip	4 Ring
2 Washer	5 Bearing
3 Spring	

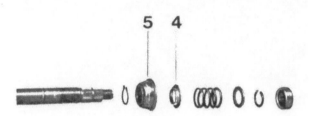

Fig. 11.35. Correct assembly of ring (4) and bearing (5) to steering shaft

22 Steering column - removal and refitting

1   Remove the shaft, as described in Section 20, and then unbolt the column lower flange plate.
2   Disconnect the ignition switch leads. Remove the dipper switch.
3   Withdraw the column upper mounting bolts (which pass through the steering column lock) and withdraw the column into the car interior.
4   Refitting is a reversal of removal.

23 Steering column lock - removal and installation

1   Remove the steering shaft (Section 20).
2   Disconnect the lead from the battery negative terminal
3   Remove the steering column trim panel.
4   Remove the dipper switch from its support bracket.
5   From the lower end of the column, unbolt the clamp bracket and the column clamp.
6   Identify the leads to the ignition switch and disconnect them.
7   Unscrew and remove the bolts which pass through the steering lock switch and serve as mounting bolts.
8   Withdraw the column complete with lock assembly.
9   Unscrew the shear bolts using a small punch or cold chisel to rotate them or alternatively, drill holes in them and use a bolt extractor.
10 Release the retaining screw and remove the lock assembly from the column.
11 Refitting and installation is a reversal of removal and dismantling but before tightening the shear bolts, make sure that the lock locating screw engages correctly in the hole drilled to receive it in the column tube. Finally tighten the shear bolts until their heads snap off.

Fig. 11.36. Steering column trim panel

Fig. 11.37. Steering column lower clamps and bracket

Fig. 11.38. Steering column upper mounting bolts

24 Steering box - removal and installation

1   Mark the relative position of the flexible coupling to the steering shaft (photo).
2   Release the two pinch bolts on the flexible coupling and push the coupling up the shaft.
3   With a suitable extractor, disconnect the relay rod balljoint from the drop arm of the steering box.
4   Turn the steering wheel to the 'straight-ahead' position and check that the marks on the drop arm and the steering roller shaft are in alignment.
5   Unbolt the steering box from the front crossmember.
6   Installation is a reversal of removal.

24.1 Steering box, shaft and flexible coupling

25 Steering box - overhaul

1   With the steering box removed from the car, prise out the oil filler plug and drain the oil.

2   Flatten the locking plate which secures the drop arm nut, unscrew the nut and pull off the drop arm with a suitable extractor noting the drop arm and sector shaft alignment marks.

3   Unbolt the steering box cover and withdraw it complete with sector shaft.

4   Remove the locknut from the cover adjustment screw and screw the adjuster out of the cover threads.

5   Remove the side cover socket screws and lift off the cover and shims; Extract the bearing track, the outer ball cage, the worm and the inner ball cage.

6   Clean all components and renew any which are worn or scored or damaged.

7   The shim pack under the side cover must be selected to provide the correct adjustment of the worm bearings when the cover is bolted down. To check this adjustment, install the worm and bearings and the original shim. Note that the cover must have its machined edge towards the bottom of the box. Wind a length of cord round the splined end of the worm and pull the cord as the worm is turning tighten the side cover bolts. If the worm seizes, remove the cover and increase the shim thickness. If the worm turns easily when the cover bolts are fully tightened, check the turning force. This should be between 3 and 4 ounces (85 to 113 g) only and unless a suitable spring balance is available install shims so that any endfloat of the worm or bearings *just* disappears. The steering box should be filled with oil for this check.

8   Turn the worm so that its mark is in alignment with the mark on the steering box then install the sector shaft so that its mark is in alignment with the steering box seam. Locate a new gasket on the rim of the steering box.

9   Engage a screwdriver in the slot in the end of the adjustment screw and turn it into the cover. Note the method of securing the adjuster screw to the sector shaft.

10 Bolt down the cover and then adjust the screw and locknut to privde an endfloat of 0.002 in (0.05 mm) on the sector shaft with the gear in the 'straight-ahead' position.

11 Always align the drop arm and sector shaft marks before fitting a new lockplate and screwing on the securing nut to the specified torque.

26 Steering idler - removal and refitting

1   Using a suitable balljoint separator, disconnect the relay rod from the idler arm (photo).

2   Disconnect the track control arm from its crossmember pivot (Section 7).

3   Remove the castellaed nut from the top of the idler housing and withdraw the idler shaft and arm assembly downwards.

4   Renew the special bearings if necessary and reassemble by reversing the dismantling process. Tighten the idler shaft castellated nut to the specified torque and insert a new split pin.

Fig. 11.40. Steering drop arm to sector shaft alignment marks

Fig. 11.41. Withdrawing steering box cover complete with sector shaft

1 Cover bolt	5 Cover
2 Cover bolt	6 Sector shaft
3 Cover bolt	
4 Cover bolt	

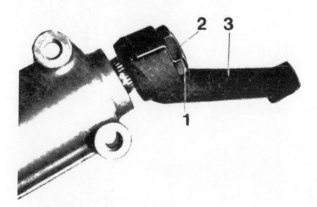

Fig. 11.39. Steering drop arm lockplate (1) nut (2) and drop arm (3)

Fig. 11.42. Steering box cover (3) and adjuster screw (2) and locknut (1)

Fig. 11.43. Steering worm components

1 Side cover
2 Shim
3 Outer bearing
4 Ball cage

5 Worm
6 Inner bearing
7 Housing

Fig. 11.46. Sector shaft adjuster screw components

1 Screw
2 Circlip

3 Shaft
4 Collar

Fig. 11.44. Steering box and worm alignment marks

26.1 Steering idler

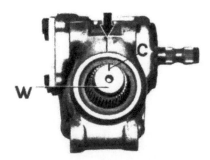

Fig. 11.45. Steering box and sector shaft alignment marks

27.2 One side of the front suspension

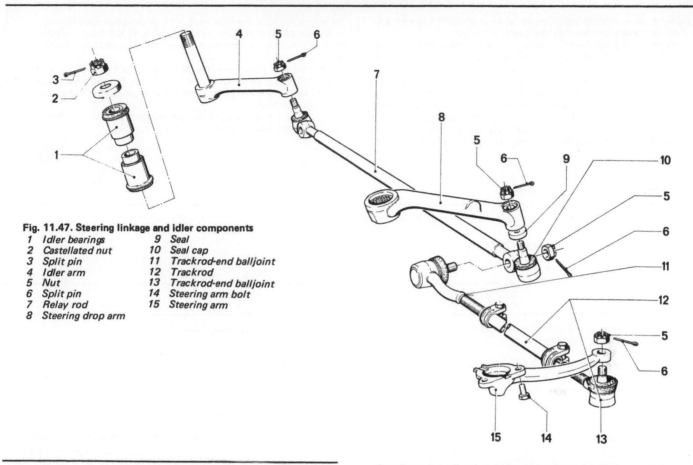

Fig. 11.47. Steering linkage and idler components

1	Idler bearings	9	Seal
2	Castellated nut	10	Seal cap
3	Split pin	11	Trackrod-end balljoint
4	Idler arm	12	Trackrod
5	Nut	13	Trackrod-end balljoint
6	Split pin	14	Steering arm bolt
7	Relay rod	15	Steering arm
8	Steering drop arm		

27 Trackrods and balljoints - removal and refitting

1   To disconnect a balljoint taper pin, first remove the castellated nut or self-locking nut then use an extractor or forked, tapered wedges to separate the balljoint from the eye of the attached component. Another method is to place the head of a hammer (or other solid metal article) on one side of the hole in the arm into which the pin is fitted. Then hit it smartly with a hammer on the opposite side. This has the effect of squeezing the taper out and usually works, provided one can get a good swing at it.

2   The balljoints at each end of the two outer track rods should be released from the steering arms and then the clamp pinch bolts released and the trackrod ends unscrewed (left and right-hand threads) from the trackrod (photo).

3   Balljoints cannot be overhauled or repaired and they should be renewed if they are worn or show any slackness in the movement of the ball within its housing. No provision is made for lubrication.

4   If new outer trackrod-ends have been fitted, set the between balljoint centres distance of the trackrods to 14.2 in (361.0 mm) as a basic setting pending front wheel alignment as described in the next Section.

28 Steering angles and front wheel alignment

1   Accurate front wheel alignment is essential for good steering and tyre wear. Before considering the steering angle, check that the tyres are correctly inflated, that the front wheels are not buckled, the hub bearings are not worn or incorrectly adjusted and that the steering linkage is in good order, without slackness or wear at the joints.

2   Wheen alignment consists of four factors:

Camber, which is the angle at which the front wheels are set from the vertical when viewed from the front of the car. Positive camber is the amount (in degrees) that the wheels are tilted outwards at the top from the vertical.

Castor is the angle between the steering axis and a vertical line when viewed from each side of the car. Positive castor is when the steering axis is inclined rearward.

Steering axis inclination is the angle, when viewed from the front of the car, between the vertical and an imaginary line drawn between the upper and lower suspension leg pivots.

Toe-in is the amount by which the distance between the front inside edges of the roadwheels (measured at hub height) is less than the diametrically opposite distance measured between the rear inside edges of the front roadwheels.

3   All steering angles other than toe-in are set in production and are not adjustable. Front wheel tracking (toe-in) checks are best carried out with modern setting equipment but a reasonably accurate alternative and adjustment procedure may be carried out as follows:

4   Place the car on level ground with the wheels in the straight ahead position.

5   Obtain or make a toe-in guage. One may be easily made from tubing, cranked to clear the sump and ballhousing, having an adjustable nut and setscrew at one end.

6   With the gauge, measure the distance between the two inner rims of the front roadwheels, at hub height and at the rear of the wheels.

7   Pull or push the vehicle so that the roadwheel turns through half a turn (180°) and measure the distance between the two inner rims at hub height at the front of the wheel. This last measurement should be less than the first by the specified toe-in (see 'Specifications' Section).

8   Where the toe-in is found to be incorrect, slacken the clamps on each outer trackrod and rotate each trackrod an equal amount but in opposite directions, until the correct toe-in is obtained. Tighten the clamps ensuring that the balljoints are held in the centre of their arc of travel during tightening.

9   Wear in the rear trailing arm bushes can cause incorrect tracking of the rear wheels with consequent heavy tyre wear.

29 Wheels and tyres

1   The roadwheels are of pressed steel type.

2   Periodically remove the wheels, clean dirt and mud from the inside and outside surfaces and examine for signs of rusting or rim damage and rectify as necessary.

3   Apply a smear of light grease to the wheel studs before screwing on the nuts and finally tighten them to specified torque.

4   The tyres fitted may be of crossply or radial construction according to territory and specification. Never mix tyres of different construction and always check and maintain the pressures regularly.

5   If the wheels have been balanced on the vehicle then it is important that the wheels are not moved round the vehicle in an effort to equalise tread wear. If a wheel is removed, then the relationship of the wheel studs to the holes in the wheel should be marked to ensure exact replacement, other the balance of wheel, hub and tyre will be upset.

6   Where the wheels have been balanced off the vehicle, then they may be moved round to equalise wear. Include the spare wheel in any rotational pattern. If radial tyres are fitted, do not move the wheels from side to side but only interchange the front and rear wheels on the same side.

7   Balancing of the wheels is an essential factor in good steering and road holding. When the tyres have been in use for about half their useful life the wheels should be rebalanced to compensate for the lost tread rubber due to wear.

8   Inspect the tyre walls and threads regularly for cuts and damage and where evident, have them professionally repaired.

30 Fault diagnosis - suspension and steering

Before diagnosing faults from the following chart, check that any irregularities are not caused by:
1   *Binding brakes*
2   *Incorrect 'mix' of radial and crossply tyres*
3   *Incorrect tyre pressures*
4   *Misalignment of the bodyframe*

Symptom	Reason/s
Steering wheel can be moved considerably before any sign of movement of the wheels is apparent	Wear in the steering linkage gear.
Vehicle difficult to steer in a consistent straight line - wandering	As above. Wheel alignment incorrect (indicated by excessive or uneven tyre wear). Front wheel hub bearings loose or worn. Worn suspension unit swivel joints.
Steering stiff and heavy	Incorrect wheel alignment (indicated by excessive or uneven tyre wear). Excessive wear or seizure in one or more of the joints in the steering linkage or suspension unit balljoints. Excessive wear in the steering gear unit.
Wheel wobble and vibration	Roadwheels out of balance. Roadwheels buckled. Wheel alignment incorrect. Wear in the steering linkage, suspension unit bearings or track control arm bushes.
Excessive pitching and rolling on corners and during braking	Defective shock absorbers and/or broken spring.

Chapter 12 Bodywork and fittings

Contents

1 General description

The body and underframe is of unitary, all-welded construction. All versions covered by this manual have two door bodywork except the 2000 Touring which has a third door (hatchback). The cabriolet has a rollbar in the interests of safety.

2 Maintenance - bodywork and underframe

1 The general condition of a car's bodywork is the one thing that significantly affects its value. Maintenance is easy but needs to be regular. Neglect, particularly after minor damage, can lead quickly to further deterioration and costly repair bills. It is important also to keep watch on those parts of the car not immediately visible, for instance the underside, inside all the wheel arches and lower part of the engine compartment.

2 The basic maintenance routine for the bodywork is washing - preferably with a lot of water, from a hose. This will remove all the loose solids which may have stuck to the car. It is important to flush these off in such a way as to prevent grit from scratching the finish.

The wheel arches and underbody need washing in the same way to remove any accumulated mud which will retain moisture and tend to encourage rust. Paradoxically enough, the best time to clean the underbody and wheel arches is in wet weather when the mud is thoroughly wet and soft. In very wet weather the underbody is usually cleaned of large accumulations automatically and this is a good time for inspection.

3 Periodically it is a good idea to have the whole of the underside of the car steam cleaned, engine compartment included, so that a thorough inspection can be carried out to see what minor repairs and renovations are necessary. Steam cleaning is available at many garages and is necessary for removal of accumulation of oily grime which sometimes is allowed to cake thick in certain areas near the engine, gearbox and back axle. If steam facilities are not available, there are one or two excellent grease solvents available which can be brush applied. The dirt can then be simply hosed off.

4 After washing paintwork, wipe off with a chamois leather to give an unspotted clear finish. A coat of clear protective wax polish will give added protection against chemical pollutants in the air. If the paintwork

sheen has dulled or oxidised, use a cleaner/polisher combination to restore the brilliance of the shine. This requires a little effort, but is usually caused because regular washing has been neglected. Always check that the door and ventilator opening drain holes and pipes are completely clear so that water can drain out. Bright work should be treated the same way as paintwork. Windscreens and windows can be kept clear of the smeary film which often appears if a little ammonia is added to the water. If they are scratched, a good rub with a proprietary metal polish will often clear them. Never use any form of wax or other body or chromium polish on glass.

3 Maintenance - upholstery and carpets

1 Mats and carpets should br brushed or vacuum cleaned regularly to keep them free of grit. If they are badly stained remove them from the car for scrubbing or sponging and make quite sure they are dry before replacement. Seats and interior trim panels can be kept clean by a wipe over with a damp cloth.

If they do become stained (which can be more apparent on light coloured upholstery) use a little liquid detergent and a soft nail brush to scour the grime out of grain of the material. Do not forget to keep the head lining clean in the same way as the upholstery. When using liquid cleaners inside the car do not over-wet the surfaces being cleaned. Excessive damp could get into the seams and padded interior causing stains, offensive odours or even rot. If the inside of the car gets wet accidentally it is worthwhile taking some trouble to dry it out properly, particularly where carpets are involved. **Do not** leave oil or electric heaters inside the car for this purpose.

4 Minor body damage - repair

See photo sequences on pages 189, 190 and 191.

Repair of minor scratches in the car's bodywork

If the scratch is very superficial, and does not penetrate to the metal of the bodywork - repair is very simple. Lightly rub the area of the scratch with a paintwork renovator (eg; T-Cut), or a very fine cutting paste, to remove loose paint from the scratch and to clear the surrounding bodywork of wax polish. Rinse the area with clean water.

Apply touch-up paint to the scratch using a thin paint brush; continue to apply thin layers of paint until the surface of the paint in the scratch is level with the surrounding paintwork. Allow the new paint at least two weeks to harden, then, blend it into the surrounding paintwork by rubbing the paintwork in the scratch area with a paintwork renovator (eg; T-Cut), or a very fine cutting paste. Finally apply wax polish.

An alternative to painting over the scratch is to use Holts Scratch-Patch. Use the same preparation for the affected area; then simply, pick a patch of a suitable size to cover the scratch completely. Hold the patch against the scratch and burnish its backing paper; the patch will adhere to the paintwork, freeing itself from the backing paper at the same time. Polish the affected area to blend the patch into the surrounding paintwork.

Where a scratch has pentrated right through to the metal of the bodywork, causing the metal to rust, a different repair technique is required. Remove any loose rust inhibiting paint (eg; Kurust) to prevent the formation of rust in the future. Using a rubber or nylon applicator fill the scratch with body-stopper paste. If required, this paste can be mixed with cellulose thinners to provide a very thin paste which is ideal for filling narrow scratches. Before the stopper paste in the scratch hardens, wrap a piece of smooth cotton rag around the tip of a finger. Dip the finger in cellulose thinners and then quickly sweep it across the surface of the stopper-paste   this will ensure that it is slightly hollowed. The scratch can now be painted over as described earlier in this Section.

Repair of dents in the car's bodywork

When deep denting of the car's bodywork has taken place, the first task is to pull the dent out, until the affected bodywork almost attains its original shape. There is little point in trying to restore the original shape completely, as the metal in the damage area will have stretched on impact and cannot be reshaped fully to its original contour. It is better to bring the level of the dent up to a point which is about 1/8 inch (3 mm) below the level of the surround bodywork. In cases where the dent is very shallow anyway, it is not worth trying to pull it out at all.

If the underside of the dent is accessible, it can be hammered out gently from behind, using a mallet with a wooden or plastic head. Whilst doing this, hold a suitable block of wood firmly against the outside of the dent. This block will absorb the impact from the hammer blows and thus prevent a large area of bodywork from being 'belled-out'.

Should the dent be in a section of the bodywork which has double skin or some other factor making it inaccessible from behind, a defferent technique is called for. Drill several small holes through the metal inside the dent area - particularly in the deeper sections. Then screw long self-tapping screws into the holes just sufficiently for them to gain a good purchase in the metal. Now the dent can be pulled out by pulling on the protruding heads of the screws with a pair of pliers.

The next stage of the repair is the removal of the paint from the damaged area, and from an inch (25.4 mm) or os of the surrounding 'sound' bodywork. This is accomplished most easily by using a wire brush or abrasive pad on a power drill, although it can be done just as effectively by hand using sheets of abrasive paper. To complete the preparations for filling, score the surface of the bare metal with a screwdriver or the tang of a file, or alternatively, drill small holes in the affected area. This will provide a really good 'key' for the filler paste.

To complete the repair see the Section on filling and respraying.

Repair of rust holes or gashes in the car's bodywork

Remove all paint from the affected area and from an inch or so of the surrounding 'sound' bodywork, using an abrasive pad or a wire brush on a power drill. If these are not available a few sheets of abrasive paper will do the job just as effectively. With the paint removed you will be able to gauge the severity of the corrosion and therefore decide whether to replace the whole panel (if this is possible) or to repair the affected area. Replacement body panels are not as expensive as most people think and it is often quicker and more satisfactory to fit a new panel than to attempt to repair large areas of corrosion.

Remove all fittings from the affected area except those which will act as a guide to the original shape of the damaged bodywork (eg; headlamp shells etc). Then, using tin snips or a hacksaw blade, remove all loose metal and any other metal badly affected by corrosion. Hammer the edges of the holes inwards in order to create a slight depression for the filler paste.

Wire brush the affected area to remove the powdery rust from the surface of the remaining metal. Paint the affected area with rust inhibiting paint; if the back of the rusted area is accessible treat this also.

Before filling can take place it will be necessary to block the hole in some way. This can be achieved by the use of one of the following materials: Zinc gauze, Aluminium tape or Polyurethane foam.

Zinc gauze is probably the best material to use for a large hole. Cut a piece to the approximate size and shape of the hole to be filled, then position it in the hole so that its edges are below the level of the surrounding bodywork. It can be retained in position by several blobs of filler paste around its periphery.

Aluminium tape should be used for small or very narrow holes. Pull a piece off the roll and trim it to the approximate size and shape required, then pull off the backing paper (if used) and stick the tape over the hole; it can be overlapped if the thickness of one piece is insufficient. Burnish down the edges of the tape with the handle of a screwdriver or similar, to ensure that the tape is securely attached to the metal underneath.

Polyurethane foam is best used where the hole is situated in a section of bodywork of complex shape, backed by a small box section (eg; where the sill panel meets the rear wheel arch - most cars). The usual mixing procedure for this foam is as follows: Put equal amounts of fluid from each of the two cans provided in the kits, into one container. Stir until the mixture begins to thicken, then quickly pour this mixture into the hole, and hold a piece of cardboard over the larger apertures. Almost immediately the polyurethane will begin to expand, gushing frantically out of any small holes left unblocked. When the foam hardens it can be cut back to just below the level of the surrounding bodywork with a hacksaw blade.

Having blocked off the hole, the affected area must now be filled and sprayed - see Section on bodywork filling and respraying.

Bodywork repairs - filling and re-spraying

Before using this Section, see the Sections on dent, deep scratch, rust hole, and gash repairs.

Many types of bodyfiller are available, but generally speaking those proprietary kits which contain a tin of filler paste and a tube of resin hardener (eg; Holts Cataloy) are best for this type of repair. A wide, flexible plastic or nylon applicator will be found invaluable for imparting a smooth and well contoured finish to the surface of the filler.

Mix up a little filler on a clean piece of card or board - use the hardener sparingly (follow the maker's instructions on the pack), otherwise the filler will set very rapidly.

Using the applicator, apply the filler paste to the prepared area; draw the applicator across the surface of the filler to achieve the correct contour and to level the filler surface. As soon as a contour that approximates the correct one is achieved, stop working the paste - if you carry on too long the paste will become sticky and begin to 'pick-up' on the applicator.

Continue to add thin layers of filler paste at twenty-minute intervals until the level of the filler is just 'proud' of the surrounding bodywork.

Once the filler has hardened, excess can be removed using a Surform plane or Dreadnought file. From then on, progressively finer grades of abrasive paper should be used, starting with a 40 grade 'wet-and-dry' paper. Always wrap the abrasive paper around a flat rubber cork, or wooden block - otherwise the surface of the filler will not be completely flat. During the smoothing of the filler surface the 'wet-and-dry' paper should be periodically rinsed in water - this will ensure that a very smooth finish is imparted to the filler at the final stage.

At this stage the 'dent' should be surrounded by a ring of bare metal, which in turn should be encircled by the finely 'feathered' edge of the good paintwork. Rinse the repair area with clean water, until all of the dust produced by the rubbing-down operating is gone.

Spray the whole repair area with a light coat of grey primer - this will show up any imperfections in the surface of the filler. Repair these imperfections with fresh filler paste or body-stopper, and once more smooth the surface with abrasive paper. If bodystopper is used, it can be mixed with cellulose thinners to form a really thin paste which is ideal for filling small holes. Repeat this spray and repair procedure until you are satisfied that the surface of the filler, and the feathered edge of the paintwork are perfect. Clean the repair area with clean water and allow to dry fully.

The repair area is now ready for spraying. Paint spraying must be carried out in a warm, dry, windless and dust free atmosphere. This condition can be created artificially if you have access to a large indoor working area, but if you are forced to work in the open, you will have

Typical example of rust damage to a body panel. Before starting ensure that you have all of the materials required to hand. The first task is to ...

... remove body fittings from effected area, except those which can act as a guide to the original shape of the damaged bodywork - the headlamp shell in this case.

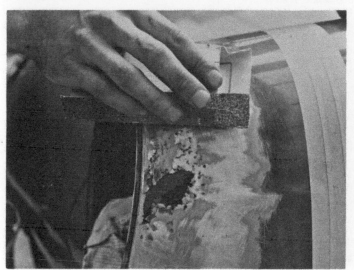

Remove all paint from the rusted area and from an inch or so of the adjoining 'sound' bodywork - use coarse abrasive paper or a power drill fitted with a wire brush or abrasive pad. Tap in the edges of the hole to provide a hollow for the filler.

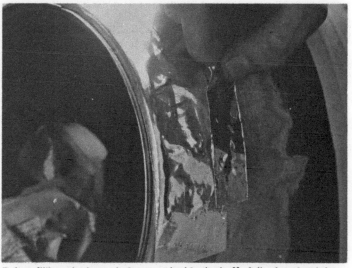

Before filling, the larger holes must be blocked off. Adhesive aluminium tape is one method; cut the tape to the required shape and size, peel off the backing strip (where used), position the tape over the hole and burnish to ensure adhesion.

Alternatively, zinc gauze can be used. Cut a piece of the gauze to the required shape and size; position it in the hole below the level of the surrounding bodywork; then ...

... secure in position by placing a few blobs of filler paste around its periphery. Alternatively, pop rivets or self-tapping screws can be used. Preparation for filling is now complete.

Mix filler and hardener according to manufacturer's instructions - avoid using too much hardener otherwise the filler will harden before you have a chance to work it.

Apply the filler to the affected area with a flexible applicator - this will ensure a smooth finish. Apply thin layers of filler at 20 minute intervals, until the surface of the filler is just 'proud' of the surrounding bodywork. Then ...

... remove excess filler and start shaping with a Surform plane or a dreadnought file. Once an approximate contour has been obtained and the surface is relatively smooth, start using ...

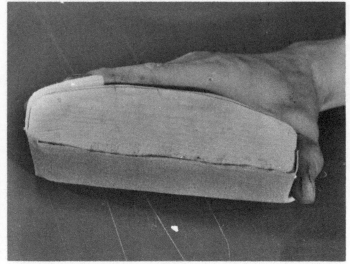

... abrasive paper. The paper should be wrapped around a flat wood, cork or rubber block - this will ensure that it imparts a smooth surface to the filler.

40 grit production paper is best to start with, then use progressively finer abrasive paper, finishing with 400 grade 'wet-and-dry'. When using 'wet-and-dry' paper, periodically rinse it in water ensuring also, that the work area is kept wet continuously.

Rubbing-down is complete when the surface of the filler is really smooth and flat, and the edges of the surrounding paintwork are finely 'feathered'. Wash the area thoroughly with clean water and allow to dry before commencing re-spray.

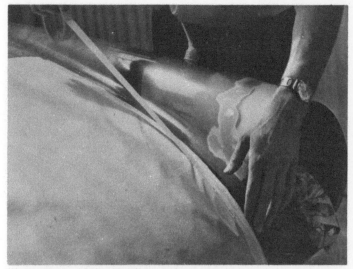

Firstly, mask off all adjoining panels and the fittings in the spray area. Ensure that the area to be sprayed is completely free of dust. Practice using an aerosol on a piece of waste metal sheet until the technique is mastered.

Spray the affected area with primer - apply several thin coats rather than one thick one. Start spraying in the centre of the repair area and then work outwards using a circular motion - in this way the paint will be evenly distributed.

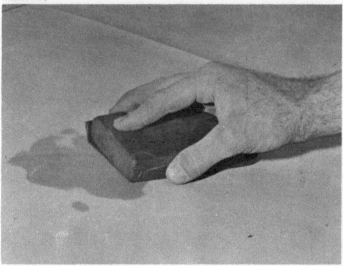

When the primer has dried inspect its surface for imperfections. Holes can be filled with filler paste or body-stopper, and lumps can be sanded smooth. Apply a further coat of primer, then 'flat' its surface with 400 grade 'wet-and-dry' paper.

Spray on the top coat, again building up the thickness with several thin coats of paint. Overspray onto the surrounding original paintwork to a depth of about five inches, applying a very thin coat at the outer edges.

Allow the new paint two weeks, at least, to harden fully, then blend it into the surrounding original paintwork with a paint restorative compound or very fine cutting paste. Use wax polish to finish off.

The finished job should look like this. Remember, the quality of the completed work is directly proportional to the amount of time and effort expended at each stage of the preparation.

to pick your day carefully. If you are working indoors, dousing the floor in the work area with water will 'lay' the dust which would otherwise be in the atmosphere. If the repair area is confined to one body panel, mask off the surrounding panels; this will help to minimise the effects of a slight mis-match in paint colours. Bodywork fittings (eg; chrome strips, door handles etc) will also need to be masked off. Use genuine masking tape and several thicknesses of newspaper for the masking operation.

Before commencing to spray, agitate the aerosol can thoroughly, then spray a test area (an old tin, or similar) until the technique is mastered. Cover the repair area with a thick coat of primer; the thickness should be built up using several thin layers of paint rather than one thick one. Using 400 grade 'wet-and-dry' paper, rub down the surface of the primer until it is really smooth. While doing this, the work area should be thoroughly doused with water, and the wet-and-dry paper periodically rinsed in water. Allow to dry before spraying on more paint.

Spray on the top coat, again building up the thickness by using several thin layers of paint. Start spraying in the centre of the repair area and then using a circular motion, work outwards until the whole repair area and about 2 inches (50 mm) of the surrounding original paintwork is covered. Remove all masking material 10 to 15 minutes after spraying on the final coat of paint. Allow the new paint at least 2 weeks to harden fully, then using a paintwork renovator (eg; T-Cut) or a very fine cutting paste, blend the edges of the new paint into the existing paintwork. Finally, apply wax polish.

5 Major body damage - repair

Where serious damage has occurred or large areas need renewal due to neglect, it means certainly that completely new sections or panels will need welding in and this is best left to professionals. If the damage is due to impact it will also be necessary to completely check the alignment of the body shell structure. Due to the principle of construction the strength and shape of the whole can be affected by damage to a part. In such instances the services of a BMW agent with specialist checking jigs are essential. If a body is left misaligned it is first of all dangerous as the car will not handle properly and secondly uneven stresses will be imposed on the steering, engine and transmission causing abnormal wear or complete failure. Tyre wear may also be excessive.

6 Maintenance - hinges and locks

1 Oil the hinges of the bonnet, boot and doors with a drop or two of light oil periodically. A good time is after the car has been washed.
2 Oil the bonnet release catch pivot pin and the safety catch pivot pin periodically.
3 Do not over lubricate door latches and strikers. Normally a little oil on the rotary cam spindle alone is sufficient.

7 Doors - tracing rattles and their rectification

1 Check first that the door is not loose at the hinges and that the latch is holding the door firmly in positin. Check also that the door lines up with the aperture in the body.
2 If the hinges are loose or the door is out of alignment it will be necessary to rest the hinge positions, as described in Section 12.
3 If the latch is holding the door properly it should hold the door tightly when fully latched and the door should line up with the body. If it is out of alignment it needs adjustment. If loose, some part of the lock mechanism must be worn out and requiring renewal.
4 Other rattle from the door would be caused by wear or looseness in the window winder, the glass channels and sill strips or the door buttons and interior latch release mechanism.

8 Door lock - removal, refitting and adjusting

1 Open the door to its fullest extent.
2 Remove the armrest retaining screws. The upper screw is accessible after prising off the cover clip and sliding it downwards (photo).
3 Remove the door lock interior handle (one screw), after prising out the plastic insert (photo).

4 Prise off the cover plate from the window regulator handle and remove the now exposed screw and pull off the handle (photo).
5 Using a piece of bent wire inserted through the small hole at the rear of the quarterlight ventilator control knob, lever off the cover plate (photos).
6 Remove the control knob screw and the knob.
7 Insert the fingers between the interior trim panel and the door and release the securing clips which are located all round the door.
8 Lift the panel away noting the two coil springs, one located on the shaft of the window regulator and the other on the quarterlight ventilator control shaft (photo).
9 Peel off the waterproof sheet from the inside of the door.
10 Remove the lock remote control assembly securing screws and unhook the control rod from the lock end.
11 Release the screw on th window glass guide channel.
12 Remove the lock securing screws from the edge of the door and then remove the lock from the door cavity in a downward direction.
13 The door exterior handle can be removd if required by unscrewing the setscrews and crosshead screw from within the door cavity.
14 Refitting is a reversal of removal but set the remote control assembly to provide a free-movement of the interior handle of between 0.112 and 0.20 in. (3.0 and 5.0 mm).
15 Make sure that the two coil springs have their larger diameters away from the handles.
16 With the lock refitted, close the door gently and check the setting of the striker plate on the door pillar. The door should close smoothly as it engages with the dovetail and peg of the striker plate and have no tendency to rattle when fully closed. If adjustment is required, release the striker plate screws and move the plate within the limits of its elongated holes in the door pillar. Do not remove these securing screws or the plate into which the screws are threaded may drop off at the back of the piller and the rear seat and body side trim panel will have to be removed to retrieve it.

9 Window glass - removal and refitting

1 Using a wedge of hardwood, tap the external weatherstrip trim upwards and remove it.
2 Remove the door interior panel, as described in the preceding Section.
3 Disconnect the glass support brackets from the glass bottom channel and the side slides (photos).
4 Support the glass and lower it about 6 in (152.4 mm) tilt the glass so that it is lower at the front of the car and lift it upwards from the door cavity.
5 Refitting is a reversal of removal but make sure that the lift arm is correctly fitted to its slot in the glass channel bracket having one plastic washer fitted each side of the bracket, the larger one being between the lift arm and the bracket.
6 Adjust the glass support brackets and side guide channels by moving them within the limits of their elongated bolt holes to ensure that the glass moves smoothly up and down without sticking or leaving gaps when fully closed.

10 Window regulator - removal and refitting

1 Remove the door window glass, as described in the preceding Section.
2 Remove the three securing bolts and withdraw the regulator mechanism from the door cavity.
3 Refitting is a reversal of removal.

11 Quarterlight (ventilator) - removal and refitting

1 Remove the door interior trim panel, as described in Section 8.
2 Release the ventilator control gear bolt and then push open the quarterlight about 1½ in (38.0 mm).
3 Push the quarterlight downwards to release it from its upper hinge and then pull it upwards and out of the door.
4 To remove the control gear, remove the three securing screws completely and extract the gear from the door cavity.
5 Refitting is a reversal of removal but if necessary, adjust the glass frame bolts so that the quarterlight has an even gap with the doorframe when the door is closed.

8.2 Armrest upper screw and cover

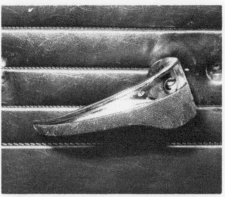

8.3 Door interior handle and cover plate

8.4 Window regulator handle and cover plate

8.5a Quarterlight (ventilator) control knob and cover plate

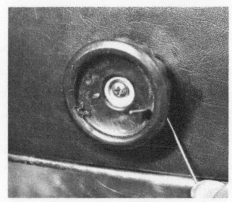

8.5b Method of removing quarterlight control knob cover plate

8.8 Window regulator control coil springs behind trim panel

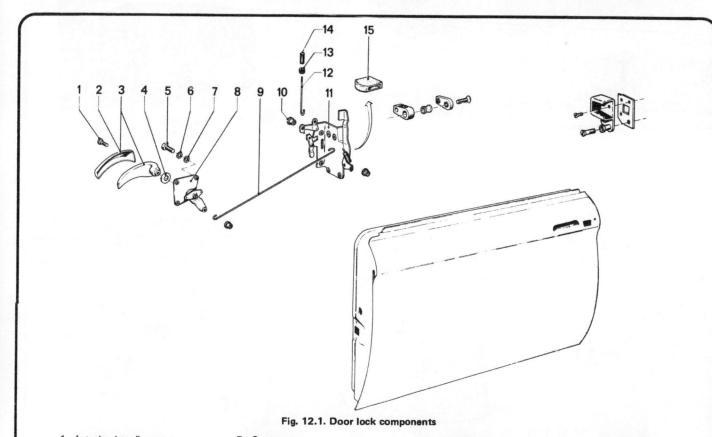

Fig. 12.1. Door lock components

1	Interior handle screw	5	Screw	9	Link rod	13	Washer
2	Handle insert	6	Lockwasher	10	Grommet	14	Knob
3	Interior handle	7	Washer	11	Lock assembly	15	Buffer
4	Washer	8	Remote control assembly	12	Lock plunger rod		

Fig. 12.2. Removing door weatherseal trim

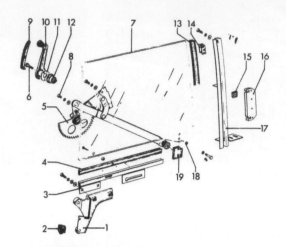

Fig. 12.3. Door glass components

1	Glass support bracket	11	Washer
2	Slide	12	Spring
3	Glass bottom channel	13	Rubber channel
4	Rubber channel	14	Clip
5	Regulator assembly	15	Slide
6	Screw	16	Glass support bracket
7	Glass	17	Guide channel
8	Regulator assembly screw	18	Bush
9	Cover plate	19	Rubber insulator
10	Regulator handle		

9.3a Door glass bottom channel

9.3b Door glass side slide

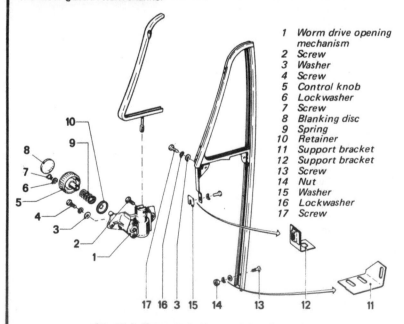

1	Worm drive opening mechanism
2	Screw
3	Washer
4	Screw
5	Control knob
6	Lockwasher
7	Screw
8	Blanking disc
9	Spring
10	Retainer
11	Support bracket
12	Support bracket
13	Screw
14	Nut
15	Washer
16	Lockwasher
17	Screw

Fig. 12.4. Quarterlight (ventilator) components

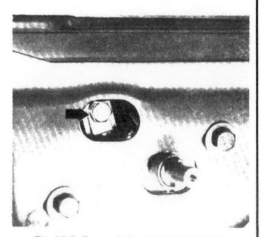

Fig. 12.5. Quarterlight control pinch bolt

12 Door - removal, refitting and adjusting

1   Remove the door interior trim panel, as described in Section 8. Disconnect the battery to prevent it discharging when the courtesy lamp switch is actuated.
2   Open the door fully and support its lower edge on jacks or blocks with some thick rag to protect the paintwork. Drive out the check strap pin.
3   Lower the window by temporarily refitting the regulator handle and after marking the position of the hinge plates on the door, release the hinge bolts.
4   Release the lower hinge bolts, having again marked their positions on the door.
5   Support the door and remove the hinge bolts and lift the door from the car body.
6   Refitting is a reversal of removal but if adjustment of the door is required to align with the body, this can be carried out by moving the door within the limits of the elongated hinge bolt holes. Finally, adjust the striker plate, as described in Section 8, paragraph 16.

13 Windscreen and rear window - removal and refitting

1   It is recommended that both these operations are left to professionals. Where the work is to be attempted however proceed in the following way.
2   Remove the wiper arms, the interior mirror or disconnect the leads from the heated rear window as appropriate.
3   Prise the bright trim from the rubber surround.
4   Run a blade round the lips of the rubber surround to ensure that it is not stuck to the body.
5   Have an assistant press one corner of the glass outwards while you pull the lip of the rubber over the body flange and restrain the glass from being ejected too violently. In the case of the windscreen, the pressure is best applied by sitting in the front seats and wearing soft soled shoes, placing the feet on the glass.
6   Unless the rubber surround is in perfect condition, renew it. Apply black mastic or sealant to the glass channel in the rubber surround and fit the surround to the glass. With the windscreen it is recommended that the bright trim is inserted into the rubber at this stage but with the rear window it is best to fit it after the glass has been installed. The use of a tool similar to the one shown will make the fitting of this trim easy.
7   Locate a length of cord in the rubber surround, in the groove which will engage with the body. Allow the ends of the cord to cross over and hang out of the groove at the bottom of the glass.
8   Offer up the glass and surround to the body aperture engaging the bottom groove. Push the glass downwards and inwards and have your assistant pull the ends of the cords evenly which will have the effect of pulling the lip of the surround over the body flange.
9   If necessary the nozzle of a mstic gun can be inserted under the outer lip of the rubber surround and a bead of sealant supplied all round to make a positive seal. Clean off any sealant according to the manufacturer's instructions and then refit the wiper arms, interior mirror or connect the heater element on the rear window.

14 Side hinged window - removal and refitting

1   Remove the screws which secure the window catch to the body at the rear edge of the glass (photo).
2   Open the window carefully (do not stain it) and peel back the rubber weatherseal at its forward (hinged) end to expose the tension type hinge plate screws (photo).
3   Remove the screws at the same time supporting the weight of the glass and then withdraw the window assembly outwards from the car.
4   Refitting is a reversal of removal.

15 Bonnet lid - removal, refitting and adjusting

1   Open the bonnet fully and mark the position of the hinge plates on the bonnet lid (photo).
2   With an assistant supporting the weight of the lid, remove the pivot bolts from the ends of the support arms on top of the front wings (photo).

3   Remove the hinge bolts and lift the bonnet away.
4   Refitting is a reversal of removal but in order to adjust the position of the bonnet it can be moved within the limits of the elongated bolt holes both on the bonnet lid or if greater adjustment is needed, remove the radiator grille and slacken the front panel hinge plate bolts which will be exposed (photo).
5   The rubber buffers can be adjusted to provide positive closure without any tendency to rattle.
6   The bonnet lock mechanism can be set and adjusted, as described in the following Section.

16 Bonnet lock - adjustment

1   Inside the car push the bonnet catch release lever against its front stop.
2   Release the clamp pinch bolt and adjust the position of the release cable until the locking bar has its end cranks horizontal and towards the front of the car. Tighten the pinch bolt.
3   To ensure positive closure, the angled catches on the underside of the bonnet lid may need slightly moving and also the rubber stops readjusting.

17 Luggage boot lid - removal, refitting and adjusting

1   Disconnect the lower ends of the luggage boot lid support struts while an assistant supports the weight of the lid.
2   Disconnect the support from the torsion rod.
3   Mark the position of the hinge plates on the underside of the boot lid and then unbolt the hinges and lift the lid from the car taking care not to damage the paintwork.
4   The torsion rods can be removed after unbolting the clamp block from the lid (photo).
5   Refitting is a reversal of removal, any adjustment of the position of the lid can be made by moving it within the limits of the elongated bolt holes.
6   Remember that if the position of the lid is altered then the lock and striker plate will need adjustment, after releasing their securing bolts.

18 Tailgate (2000 Touring-Hatchback) - removal, refitting and adjusting

1   Open the tailgate to its fullest extent and have an assistant support its weight.
2   Disconnect the leads from the window heater element.
3   Disconnect the support struts from the tailgate. Note that these struts are gas-filled and must not be dismantled or punctured, only renewed as sealed units.
4   Mark the position of the hinge plates on the body and then unscrew the hinge screws and lift the tailgate away.
5   When refitting, align the tailgate by moving it if necessary within the limits of the elongated bolt holes on the body. Adjust the lock, striker plate and rubber buffers to provide positive closure without rattling.

19 Bumpers - removal and refitting

1   Removal of the front and rear bumpers is simply a matter of unscrewing the retaining nuts and bolts.
2   In the case of rear bumpers access to the securing bolts is obtained after lifting the luggage compartment floor.
3   Disconnect the ear licence plate lamp before removing the bumper.

20 Radiator grille - removal and refitting

1   The outer sections of the radiator grille are removed after unscrewing the self-tapping screws (photo).
2   The centre section is retained by nuts and these are accessible after the outer sections of the grille have been removed.

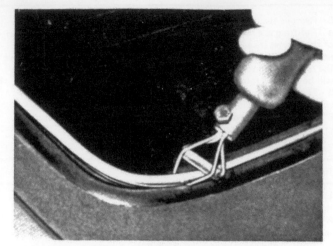

Fig. 12.6. Fitting trim to windscreen rubber surround

Fig. 12.7. Bonnet release control lever

Fig. 12.8. Bonnet release cable pinch bolt (1)

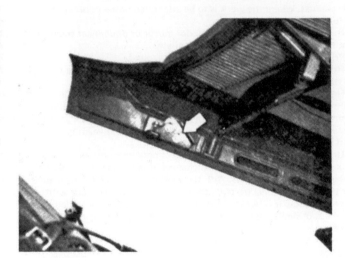

Fig. 12.9. Bonnet lock catches on underside of lid

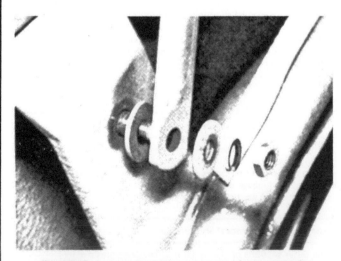

Fig. 12.10. Luggage boot lid support strut lower mounting

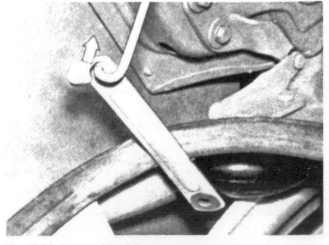

Fig. 12.11. Luggage boot lid support strut to torsion rod connection

14.1 Rear opening window catch

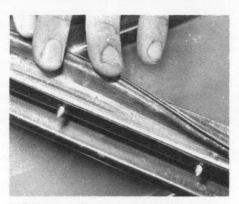
14.2 Rear opening window hinge and weatherseal

15.1 Bonnet hinge plate

15.2 Bonnet stay pivot bolt

15.4 Bonnet hinge lower plate and location of horn

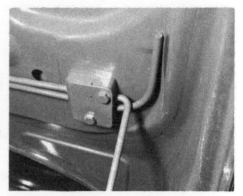
17.4 Luggage boot lid torsion rod clamp

Fig. 12.12. Tailgate hinge and support strut on 2000 Touring

21 Rear view mirrors - removal

1 The interior mirror is removed by pulling it towards the rear of the car and at the same time bending it upwards.
2 The external mirror mounted on the door is either retained by two exposed screws or a single screw covered by a rubber cover, depending on model.

22 Glove compartment - removal

1 Remove the covering trim panel.
2 Release the hinge securing screws and pull the glove compartment towards the rear of the car.

23 Rear seat (Saloon models) - removal and refitting

1 Place the fingers under the bottom edge of the seat and pull it forwards and upwards.
2 Remove the screws from the retaining straps at the base of the backrest and lift the backrest upwards.
3 Refitting is a reversal of removal.

24 Rear seat (2000 Touring) - removal and refitting

1 Release the seat catch and fold the seat fully forward.
2 Disconnect the spring from the cover flap and pull out the rubber buffer.
3 Extract the circlips (noting the location of the washers) from the pivot pins. Withdraw the pivot pins and remove the seat back from its mountings.
4 Disconnect the spring from the lever on the seat mounting and remove the pivot pin and its nut.
5 Raise the seat at its rear edge, remove the side mountings and withdraw the seat.
6 Refitting is a reversal of removal but adjust the height of the seat back buffer so that it just makes contact when the seat back has been locked in position.

25 Sliding roof - adjustment

1 If a sliding roof is fitted and it has gone out of adjustment, first open the sliding roof by about 8 in (203 mm) and then prise down the headlining at the front of the roof.
2 Remove the sliding roof regulator handle and turn it in a clockwise direction until it reaches its stop then turn it back two complete turns.
3 Close the sliding roof squarely to its front edge by pushing it with the hands.

4   Refit the regulator handle and open and close the roof several times.

5   With the roof in the closed position, again remove the regulator handle and turn it to its stop. Refit the regulator and set the knob in the centre of the knob recess.

6   Ideally, the sliding roof should be set so that its surface is 0.04 in (1.0 mm) lower at the front and a similar amount higher at the rear than the adjacent roof outer skin. Rear adjustment can be carried out by releasing the screw on the serrated adjuster. Adjustment of the level at the front of the sliding roof can be made by turning the knurled screws on the guide rail forks.

26 Heater and ventilation system - description

1   A fresh air type heater is installed which gives a very wide choice of temperature and airflow pattern according to the setting of the fascia panel control levers.

2   The heat for the system comes from the engine coolant and fresh air is admitted through the grille just below the windscreen while stale air is exhausted through the outlets hidden by the edges of the luggage compartment lid. On 2000 Touring models, the outlets are located on each side at the rear of the car and covered by small grilles (photo).

27 Heater - removal and installation

All models, except 2000 Touring and 1971 models

1   Disconnect the lead from the battery negative terminal.

2   Move the heater control lever to its hottest position and drain the cooling system.

3   Disconnect the heater return hose from its location adjacent to the windscreen wiper motor at the rear of the engine compartment (photo).

4   Disconnect the heater inlet hose from the water valve which is located in the cavity at the engine compartment rear bulkhead (photo).

5   Working within the car carry out the following operations:
   a) *Remove the centre console (Chapter 10, Section 24).*
   b) *Remove the trim panel which covers the steering column.*
   c) *Remove the trim panels from below the fascia panel.*
   d) *Remove the steering column upper shroud.*

6   Pull the two knobs from the heater air distribution controls on the left-hand side of the steering column and then remove the control escutcheon panel (two external screws and two inner knurled nuts).

7   Pull out the ashtray, pull the knob from the heater control lever and remove the control escutcheon panel and the inner control screws.

8   Disconnect the heater electrical leads.

9   Unscrew and remove the two heater mounting nuts.

10 One is located within the glove compartment and the other on the rear of the engine bulkhead.

11 Disconnect the demister hose so that when unbolted, the bracket at the base of the steering column can be slightly turned.

12 Remove the narrow trim panel from beneath the glove compartment.

13 Disconnect the remaining demister hose and lift out the heater assembly.

14 Installation is a reversal of removal but fill the cooling system as described in Chapter 2.

On 2000 Touring and 1971 models

15 The operations are similar to those described previously for other models, but the location of the mounting nuts should be observed and the securing of the earth lead below the left-hand nut. Make sure that on installation, a new heater sealing gasket is used and the foam rubber block returned to its original position.

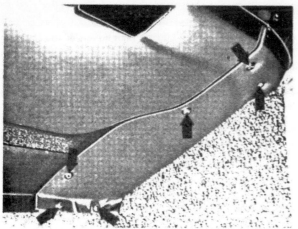

Fig. 12.13. Glove compartment hinge covering trim panel

Fig. 12.14. Rear seat (2000 Touring) spring (1) and rubber buffer (2)

20.1 Removing a section of the radiator grille

26.2 Air extraction outlet

27.4 Location of heater water valve

Fig. 12.15. Rear seat pivot circlip (2000 Touring)

Fig. 12.16. Seat mounting spring and pivot (2000 Touring)

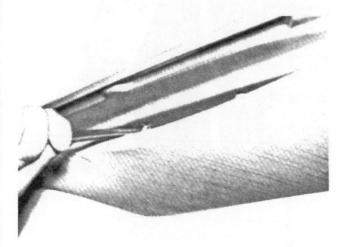

Fig. 12.17. Prising down headlining from sliding roof panel

Fig. 12.18. Sliding roof rear adjustment screw (1)

Fig. 12.19. Sliding roof front adjustment screws (2)

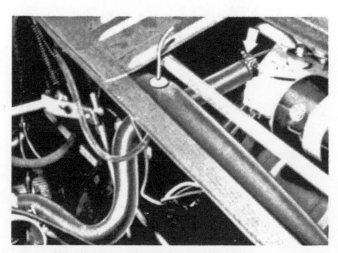

Fig. 12.20. Location of heater return hose

Fig. 12.21. Location of heater mounting nut (bulkhead) - saloon

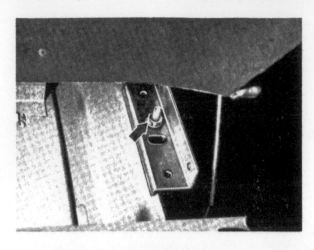

Fig. 12.22. Location of heater mounting nut (glove compartment) - saloon

Fig. 12.23. Heater mounting nut and connecting leads (2000 Touring and 1971 models)

Fig. 12.24. Heater mounting nut (bulkhead) - 2000 Touring and 1971 models

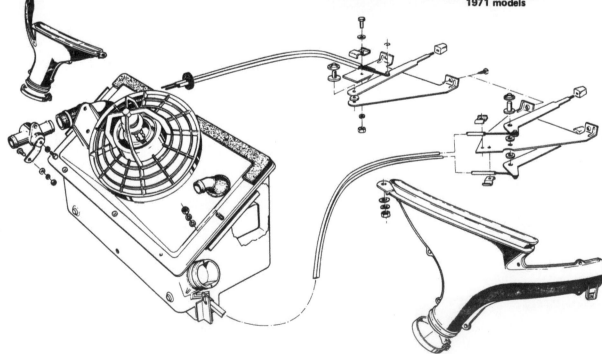

Fig. 12.25. Heater components

28 Heater - overhaul

1   The heater must be removed from the car before any dismantling can be carried out.

2   If the matrix is blocked with sediment, try reverse flushing or the use of a proprietary cleanser to clear it. If it is leaking it is recommended that it is renewed as the repair of a matrix is a skilled job.

3   If the motor is faulty it can be removed by drilling out the securing rivets and removing the clamps from the heater housing and then releasing the water valve and the rubber sleeves. The motor and fan must not be separated and have been balanced in production. Renew them if necessary as an assembly.

29 Fault diagnosis - heater

Symptom	Reason/s
Insufficient heat	Faulty or incorrect type cooling system thermostat. Coolant level too low. Faulty heater cock. Faulty ventilator valve.
Insufficient airflow	Ventilator or heat valve not operating correctly. Blower speed too low or non-existent due to blown fuse.
Faulty air deflection or temperature generally	Incorrectly adjusted cables. Disconnected demister hoses.

General repair procedures

Whenever servicing, repair or overhaul work is carried out on the car or its components, it is necessary to observe the following procedures and instructions. This will assist in carrying out the operation efficiently and to a professional standard of workmanship.

Joint mating faces and gaskets

Where a gasket is used between the mating faces of two components, ensure that it is renewed on reassembly, and fit it dry unless otherwise stated in the repair procedure. Make sure that the mating faces are clean and dry with all traces of old gasket removed. When cleaning a joint face, use a tool which is not likely to score or damage the face, and remove any burrs or nicks with an oilstone or fine file.

Make sure that tapped holes are cleaned with a pipe cleaner, and keep them free of jointing compound if this is being used unless specifically instructed otherwise.

Ensure that all orifices, channels or pipes are clear and blow through them, preferably using compressed air.

Oil seals

Whenever an oil seal is removed from its working location, either individually or as part of an assembly, it should be renewed.

The very fine sealing lip of the seal is easily damaged and will not seal if the surface it contacts is not completely clean and free from scratches, nicks or grooves. If the original sealing surface of the component cannot be restored, the component should be renewed.

Protect the lips of the seal from any surface which may damage them in the course of fitting. Use tape or a conical sleeve where possible. Lubricate the seal lips with oil before fitting and, on dual lipped seals, fill the space between the lips with grease.

Unless otherwise stated, oil seals must be fitted with their sealing lips toward the lubricant to be sealed.

Use a tubular drift or block of wood of the appropriate size to install the seal and, if the seal housing is shouldered, drive the seal down to the shoulder. If the seal housing is unshouldered, the seal should be fitted with its face flush with the housing top face.

Screw threads and fastenings

Always ensure that a blind tapped hole is completely free from oil, grease, water or other fluid before installing the bolt or stud. Failure to do this could cause the housing to crack due to the hydraulic action of the bolt or stud as it is screwed in.

When tightening a castellated nut to accept a split pin, tighten the nut to the specified torque, where applicable, and then tighten further to the next split pin hole. Never slacken the nut to align a split pin hole unless stated in the repair procedure.

When checking or retightening a nut or bolt to a specified torque setting, slacken the nut or bolt by a quarter of a turn, and then retighten to the specified setting.

Locknuts, locktabs and washers

Any fastening which will rotate against a component or housing in the course of tightening should always have a washer between it and the relevant component or housing.

Spring or split washers should always be renewed when they are used to lock a critical component such as a big-end bearing retaining nut or bolt.

Locktabs which are folded over to retain a nut or bolt should always be renewed.

Self-locking nuts can be reused in non-critical areas, providing resistance can be felt when the locking portion passes over the bolt or stud thread.

Split pins must always be replaced with new ones of the correct size for the hole.

Special tools

Some repair procedures in this manual entail the use of special tools such as a press, two or three-legged pullers, spring compressors etc. Wherever possible, suitable readily available alternatives to the manufacturer's special tools are described, and are shown in use. In some instances, where no alternative is possible, it has been necessary to resort to the use of a manufacturer's tool and this has been done for reasons of safety as well as the efficient completion of the repair operation. Unless you are highly skilled and have a thorough understanding of the procedure described, never attempt to bypass the use of any special tool when the procedure described specifies its use. Not only is there a very great risk of personal injury, but expensive damage could be caused to the components involved.

Safety first!

Regardless of how enthusiastic you may be about getting on with the job at hand, take the time to ensure that your safety is not jeopardized. A moment's lack of attention can result in an accident, as can failure to observe certain simple safety precautions. The possibility of an accident will always exist, and the following points should not be considered a comprehensive list of all dangers. Rather, they are intended to make you aware of the risks and to encourage a safety conscious approach to all work you carry out on your vehicle.

Essential DOs and DON'Ts

DON'T rely on a jack when working under the vehicle. Always use approved jackstands to support the weight of the vehicle and place them under the recommended lift or support points.

DON'T attempt to loosen extremely tight fasteners (i.e. wheel lug nuts) while the vehicle is on a jack — it may fall.

DON'T start the engine without first making sure that the transmission is in Neutral (or Park where applicable) and the parking brake is set.

DON'T remove the radiator cap from a hot cooling system — let it cool or cover it with a cloth and release the pressure gradually.

DON'T attempt to drain the engine oil until you are sure it has cooled to the point that it will not burn you.

DON'T touch any part of the engine or exhaust system until it has cooled sufficiently to avoid burns.

DON'T siphon toxic liquids such as gasoline, antifreeze and brake fluid by mouth, or allow them to remain on your skin.

DON'T inhale brake lining dust — it is potentially hazardous (see *Asbestos* below)

DON'T allow spilled oil or grease to remain on the floor — wipe it up before someone slips on it.

DON'T use loose fitting wrenches or other tools which may slip and cause injury.

DON'T push on wrenches when loosening or tightening nuts or bolts. Always try to pull the wrench toward you. If the situation calls for pushing the wrench away, push with an open hand to avoid scraped knuckles if the wrench should slip.

DON'T attempt to lift a heavy component alone — get someone to help you.

DON'T rush or take unsafe shortcuts to finish a job.

DON'T allow children or animals in or around the vehicle while you are working on it.

DO wear eye protection when using power tools such as a drill, sander, bench grinder, etc. and when working under a vehicle.

DO keep loose clothing and long hair well out of the way of moving parts.

DO make sure that any hoist used has a safe working load rating adequate for the job.

DO get someone to check on you periodically when working alone on a vehicle.

DO carry out work in a logical sequence and make sure that everything is correctly assembled and tightened.

DO keep chemicals and fluids tightly capped and out of the reach of children and pets.

DO remember that your vehicle's safety affects that of yourself and others. If in doubt on any point, get professional advice.

Asbestos

Certain friction, insulating, sealing, and other products — such as brake linings, brake bands, clutch linings, torque converters, gaskets, etc. — contain asbestos. *Extreme care must be taken to avoid inhalation of dust from such products since it is hazardous to health.* If in doubt, assume that they *do* contain asbestos.

Fire

Remember at all times that gasoline is highly flammable. Never smoke or have any kind of open flame around when working on a vehicle. But the risk does not end there. A spark caused by an electrical short circuit, by two metal surfaces contacting each other, or even by static electricity built up in your body under certain conditions, can ignite gasoline vapors, which in a confined space are highly explosive. Do not, under any circumstances, use gasoline for cleaning parts. Use an approved safety solvent.

Always disconnect the battery ground (–) cable *at the battery* before working on any part of the fuel system or electrical system. Never risk spilling fuel on a hot engine or exhaust component.

It is strongly recommended that a fire extinguisher suitable for use on fuel and electrical fires be kept handy in the garage or workshop at all times. Never try to extinguish a fuel or electrical fire with water.

Torch (flashlight in the US)

Any reference to a "torch" appearing in this manual should always be taken to mean a hand-held, battery-operated electric light or flashlight. It DOES NOT mean a welding or propane torch or blowtorch.

Fumes

Certain fumes are highly toxic and can quickly cause unconsciousness and even death if inhaled to any extent. Gasoline vapor falls into this category, as do the vapors from some cleaning solvents. Any draining or pouring of such volatile fluids should be done in a well ventilated area.

When using cleaning fluids and solvents, read the instructions on the container carefully. Never use materials from unmarked containers.

Never run the engine in an enclosed space, such as a garage. Exhaust fumes contain carbon monoxide, which is extremely poisonous. If you need to run the engine, always do so in the open air, or at least have the rear of the vehicle outside the work area.

If you are fortunate enough to have the use of an inspection pit, never drain or pour gasoline and never run the engine while the vehicle is over the pit. The fumes, being heavier than air, will concentrate in the pit with possibly lethal results.

The battery

Never create a spark or allow a bare light bulb near a battery. They normally give off a certain amount of hydrogen gas, which is highly explosive.

Always disconnect the battery ground (–) cable *at the battery* before working on the fuel or electrical systems.

If possible, loosen the filler caps or cover when charging the battery from an external source (this does not apply to sealed or maintenance-free batteries). Do not charge at an excessive rate or the battery may burst.

Take care when adding water to a non maintenance-free battery and when carrying a battery. The electrolyte, even when diluted, is very corrosive and should not be allowed to contact clothing or skin.

Always wear eye protection when cleaning the battery to prevent the caustic deposits from entering your eyes.

Mains electricity (household current in the US)

When using an electric power tool, inspection light, etc., which operates on household current, always make sure that the tool is correctly connected to its plug and that, where necessary, it is properly grounded. Do not use such items in damp conditions and, again, do not create a spark or apply excessive heat in the vicinity of fuel or fuel vapor.

Secondary ignition system voltage

A severe electric shock can result from touching certain parts of the ignition system (such as the spark plug wires) when the engine is running or being cranked, particularly if components are damp or the insulation is defective. In the case of an electronic ignition system, the secondary system voltage is much higher and could prove fatal.

Conversion factors

Length (distance)

Inches (in)	X 25.4	= Millimetres (mm)	X 0.0394	= Inches (in)
Feet (ft)	X 0.305	= Metres (m)	X 3.281	= Feet (ft)
Miles	X 1.609	= Kilometres (km)	X 0.621	= Miles

Volume (capacity)

Cubic inches (cu in; in³)	X 16.387	= Cubic centimetres (cc; cm³)	X 0.061	= Cubic inches (cu in; in³)
Imperial pints (Imp pt)	X 0.568	= Litres (l)	X 1.76	= Imperial pints (Imp pt)
Imperial quarts (Imp qt)	X 1.137	= Litres (l)	X 0.88	= Imperial quarts (Imp qt)
Imperial quarts (Imp qt)	X 1.201	= US quarts (US qt)	X 0.833	= Imperial quarts (Imp qt)
US quarts (US qt)	X 0.946	= Litres (l)	X 1.057	= US quarts (US qt)
Imperial gallons (Imp gal)	X 4.546	= Litres (l)	X 0.22	= Imperial gallons (Imp gal)
Imperial gallons (Imp gal)	X 1.201	= US gallons (US gal)	X 0.833	= Imperial gallons (Imp gal)
US gallons (US gal)	X 3.785	= Litres (l)	X 0.264	= US gallons (US gal)

Mass (weight)

Ounces (oz)	X 28.35	= Grams (g)	X 0.035	= Ounces (oz)
Pounds (lb)	X 0.454	= Kilograms (kg)	X 2.205	= Pounds (lb)

Force

Ounces-force (ozf; oz)	X 0.278	= Newtons (N)	X 3.6	= Ounces-force (ozf; oz)
Pounds-force (lbf; lb)	X 4.448	= Newtons (N)	X 0.225	= Pounds-force (lbf; lb)
Newtons (N)	X 0.1	= Kilograms-force (kgf; kg)	X 9.81	= Newtons (N)

Pressure

Pounds-force per square inch (psi; lbf/in²; lb/in²)	X 0.070	= Kilograms-force per square centimetre (kgf/cm²; kg/cm²)	X 14.223	= Pounds-force per square inch (psi; lbf/in²; lb/in²)
Pounds-force per square inch (psi; lbf/in²; lb/in²)	X 0.068	= Atmospheres (atm)	X 14.696	= Pounds-force per square inch (psi; lbf/in²; lb/in²)
Pounds-force per square inch (psi; lbf/in²; lb/in²)	X 0.069	= Bars	X 14.5	= Pounds-force per square inch (psi; lbf/in²; lb/in²)
Pounds-force per square inch (psi; lbf/in²; lb/in²)	X 6.895	= Kilopascals (kPa)	X 0.145	= Pounds-force per square inch (psi; lbf/in²; lb/in²)
Kilopascals (kPa)	X 0.01	= Kilograms-force per square centimetre (kgf/cm²; kg/cm²)	X 98.1	= Kilopascals (kPa)
Millibar (mbar)	X 100	= Pascals (Pa)	X 0.01	= Millibar (mbar)
Millibar (mbar)	X 0.0145	= Pounds-force per square inch (psi; lbf/in²; lb/in²)	X 68.947	= Millibar (mbar)
Millibar (mbar)	X 0.75	= Millimetres of mercury (mmHg)	X 1.333	= Millibar (mbar)
Millibar (mbar)	X 0.401	= Inches of water (inH₂O)	X 2.491	= Millibar (mbar)
Millimetres of mercury (mmHg)	X 0.535	= Inches of water (inH₂O)	X 1.868	= Millimetres of mercury (mmHg)
Inches of water (inH₂O)	X 0.036	= Pounds-force per square inch (psi; lbf/in²; lb/in²)	X 27.68	= Inches of water (inH₂O)

Torque (moment of force)

Pounds-force inches (lbf in; lb in)	X 1.152	= Kilograms-force centimetre (kgf cm; kg cm)	X 0.868	= Pounds-force inches (lbf in; lb in)
Pounds-force inches (lbf in; lb in)	X 0.113	= Newton metres (Nm)	X 8.85	= Pounds-force inches (lbf in; lb in)
Pounds-force inches (lbf in; lb in)	X 0.083	= Pounds-force feet (lbf ft; lb ft)	X 12	= Pounds-force inches (lbf in; lb in)
Pounds-force feet (lbf ft; lb ft)	X 0.138	= Kilograms-force metres (kgf m; kg m)	X 7.233	= Pounds-force feet (lbf ft; lb ft)
Pounds-force feet (lbf ft; lb ft)	X 1.356	= Newton metres (Nm)	X 0.738	= Pounds-force feet (lbf ft; lb ft)
Newton metres (Nm)	X 0.102	= Kilograms-force metres (kgf m; kg m)	X 9.804	= Newton metres (Nm)

Power

Horsepower (hp)	X 745.7	= Watts (W)	X 0.0013	= Horsepower (hp)

Velocity (speed)

Miles per hour (miles/hr; mph)	X 1.609	= Kilometres per hour (km/hr; kph)	X 0.621	= Miles per hour (miles/hr; mph)

Fuel consumption*

Miles per gallon, Imperial (mpg)	X 0.354	= Kilometres per litre (km/l)	X 2.825	= Miles per gallon, Imperial (mpg)
Miles per gallon, US (mpg)	X 0.425	= Kilometres per litre (km/l)	X 2.352	= Miles per gallon, US (mpg)

Temperature

Degrees Fahrenheit = (°C x 1.8) + 32

Degrees Celsius (Degrees Centigrade; °C) = (°F - 32) x 0.56

*It is common practice to convert from miles per gallon (mpg) to litres/100 kilometres (l/100km), where mpg (Imperial) x l/100 km = 282 and mpg (US) x l/100 km = 235

Index

Haynes Automotive Manuals

NOTE: If you do not see a listing for your vehicle, please visit our website **haynes.com** *for the latest product information.*

ACURA
12020 Integra '86 thru '89 & Legend '86 thru '90
12021 Integra '90 thru '93 & Legend '91 thru '95
Integra '94 thru '00 - *see HONDA Civic (42025)*
MDX '01 thru '07 - *see HONDA Pilot (42037)*
12050 Acura TL all models '99 thru '08

AMC
Jeep CJ - *see JEEP (50020)*
14020 Concord/Hornet/Gremlin/Spirit '70 thru '83
14025 (Renault) Alliance & Encore '83 thru '87

AUDI
15020 4000 all models '80 thru '87
15025 5000 all models '77 thru '83
15026 5000 all models '84 thru '88
Audi A4 '96 thru '01 - *see VW Passat (96023)*
15030 Audi A4 '02 thru '08

AUSTIN
Healey Sprite - *see MG Midget (66015)*

BMW
18020 3/5 Series '82 thru '92
18021 3 Series including Z3 models '92 thru '98
18022 3-Series incl. Z4 models '99 thru '05
18023 3-Series '06 thru '10
18025 320i all 4 cyl models '75 thru '83
18050 1500 thru 2002 except Turbo '59 thru '77

BUICK
19010 Buick Century '97 thru '05
Century (front-wheel drive) - *see GM (38005)*
19020 Buick, Oldsmobile & Pontiac Full-size (Front wheel drive) '85 thru '05
19025 Buick, Oldsmobile & Pontiac Full-size (Rear wheel drive) '70 thru '90
19030 Mid-size Regal & Century '74 thru '87
Regal - *see GENERAL MOTORS (38010)*
Skyhawk - *see GM (38030)*
Skylark - *see GM (38020, 38025)*
Somerset - *see GENERAL MOTORS (38025)*

CADILLAC
21015 CTS & CTS-V '03 thru '12
21030 Cadillac Rear Wheel Drive '70 thru '93
Cimarron, Eldorado & Seville - *see GM (38015, 38030, 38031)*

CHEVROLET
10305 Chevrolet Engine Overhaul Manual
24010 Astro & GMC Safari Mini-vans '85 thru '05
24015 Camaro V8 all models '70 thru '81
24016 Camaro all models '82 thru '92
Cavalier - *see GM (38015)*
Celebrity - *see GM (38005)*
24017 Camaro & Firebird '93 thru '02
24020 Chevelle, Malibu, El Camino '69 thru '87
24024 Chevette & Pontiac T1000 '76 thru '87
Citation - *see GENERAL MOTORS (38020)*
24027 Colorado & GMC Canyon '04 thru '10
24032 Corsica/Beretta all models '87 thru '96
24040 Corvette all V8 models '68 thru '82
24041 Corvette all models '84 thru '96
24045 Full-size Sedans Caprice, Impala, Biscayne, Bel Air & Wagons '69 thru '90
24046 Impala SS & Caprice and Buick Roadmaster '91 thru '96
Impala '00 thru '05 - *see LUMINA (24048)*
24047 Impala & Monte Carlo all models '06 thru '11
Lumina '90 thru '94 - *see GM (38010)*
24048 Lumina & Monte Carlo '95 thru '05
Lumina APV - *see GM (38035)*
24050 Luv Pick-up all 2WD & 4WD '72 thru '82
Malibu - *see GM (38026)*
24055 Monte Carlo all models '70 thru '88
Monte Carlo '95 thru '01 - *see LUMINA*
24059 Nova all V8 models '69 thru '79
24060 Nova/Geo Prizm '85 thru '92
24064 Pick-ups '67 thru '87 - Chevrolet & GMC
24065 Pick-ups '88 thru '98 - Chevrolet & GMC
24066 Pick-ups '99 thru '06 - Chevrolet & GMC
24067 Chevy Silverado & GMC Sierra '07 thru '12
24070 S-10 & GMC S-15 Pick-ups '82 thru '93
24071 S-10, Sonoma & Jimmy '94 thru '04
24072 Chevrolet TrailBlazer, GMC Envoy & Oldsmobile Bravada '02 thru '09
24075 Sprint '85 thru '88, Geo Metro '89 thru '01
24080 Vans - Chevrolet & GMC '68 thru '96
24081 Full-size Vans '96 thru '10

CHRYSLER
10310 Chrysler Engine Overhaul Manual
25015 Chrysler Cirrus, Dodge Stratus, Plymouth Breeze, '95 thru '00
25020 Full-size Front-Wheel Drive '88 thru '93
K-Cars - *see DODGE Aries (30008)*
Laser - *see DODGE Daytona (30030)*
25025 Chrysler LHS, Concorde, New Yorker, Dodge Intrepid, Eagle Vision, '93 thru '97
25026 Chrysler LHS, Concorde, 300M, Dodge Intrepid '98 thru '04
25027 Chrysler 300, Dodge Charger & Magnum '05 thru '09
25030 Chrysler/Plym. Mid-size '82 thru '95
Rear-wheel Drive - *see DODGE (30050)*
25035 PT Cruiser all models '01 thru '10
25040 Chrysler Sebring '95 thru '06, Dodge Stratus '01 thru '06, Dodge Avenger '95 thru '00

DATSUN
28005 200SX all models '80 thru '83
28007 B-210 all models '73 thru '78
28009 210 all models '78 thru '82
28012 240Z, 260Z & 280Z Coupe '70 thru '78
28014 280ZX Coupe & 2+2 '79 thru '83
300ZX - *see NISSAN (72010)*
28018 510 & PL521 Pick-up '68 thru '73
28020 510 all models '78 thru '81
28022 620 Series Pick-up all models '73 thru '79
720 Series Pick-up - *see NISSAN (72030)*
28025 810/Maxima all gas models '77 thru '84

DODGE
400 & 600 - *see CHRYSLER (25030)*
30008 Aries & Plymouth Reliant '81 thru '89
30010 Caravan & Ply. Voyager '84 thru '95
30011 Caravan & Ply. Voyager '96 thru '02
30012 Challenger/Plymouth Saporro '78 thru '83
Challenger '67-'76 - *see DART (30025)*
30013 Caravan, Chrysler Voyager, Town & Country '03 thru '07
30016 Colt/Plymouth Champ '78 thru '87
30020 Dakota Pick-ups all models '87 thru '96
30021 Durango '98 & '99, Dakota '97 thru '99
30022 Durango '00 thru '03, Dakota '00 thru '04
30023 Durango '04 thru '09, Dakota '05 thru '11
30025 Dart, Challenger/Plymouth Barracuda & Valiant 6 cyl models '67 thru '76
30030 Daytona & Chrysler Laser '84 thru '89
Intrepid - *see Chrysler (25025, 25026)*
30034 Dodge & Plymouth Neon '95 thru '99
30035 Omni & Plymouth Horizon '78 thru '90
30036 Dodge and Plymouth Neon '00 thru '05
30040 Pick-ups all full-size models '74 thru '93
30041 Pick-ups all full-size models '94 thru '01
30042 Pick-ups full-size models '02 thru '08
30045 Ram 50/D50 Pick-ups & Raider and Plymouth Arrow Pick-ups '79 thru '93
30050 Dodge/Ply./Chrysler RWD '71 thru '89
30055 Shadow/Plymouth Sundance '87 thru '94
30060 Spirit & Plymouth Acclaim '89 thru '95
30065 Vans - Dodge & Plymouth '71 thru '03

EAGLE
Talon - *see MITSUBISHI (68030, 68031)*
Vision - *see CHRYSLER (25025)*

FIAT
34010 124 Sport Coupe & Spider '68 thru '78
34025 X1/9 all models '74 thru '80

FORD
10320 Ford Engine Overhaul Manual
10355 Ford Automatic Transmission Overhaul
11500 Mustang '64-1/2 thru '70 Restoration Guide
36004 Aerostar Mini-vans '86 thru '97
Aspire - *see FORD Festiva (36030)*
36006 Contour/Mercury Mystique '95 thru '00
36008 Courier Pick-up all models '72 thru '82
36012 Crown Victoria & Mercury Grand Marquis '88 thru '10
36016 Escort/Mercury Lynx '81 thru '90
36020 Escort/Mercury Tracer '91 thru '02
Expedition - *see FORD Pick-up (36059)*
36022 Escape & Mazda Tribute '01 thru '11
36024 Explorer & Mazda Navajo '91 thru '01
36025 Explorer/Mercury Mountaineer '02 thru '10
36028 Fairmont & Mercury Zephyr '78 thru '83
36030 Festiva & Aspire '88 thru '97
36032 Fiesta all models '77 thru '80
36034 Focus all models '00 thru '11
36036 Ford & Mercury Full-size '75 thru '87
36044 Ford & Mercury Mid-size '75 thru '86
36045 Ford Fusion & Mercury Milan '06 thru '10
36048 Mustang V8 all models '64-1/2 thru '73
36049 Mustang II 4 cyl, V6 & V8 '74 thru '78
36050 Mustang & Mercury Capri '79 thru '93
36051 Mustang all models '94 thru '04
36052 Mustang '05 thru '10
36054 Pick-ups and Bronco '73 thru '79
36058 Pick-ups and Bronco '80 thru '96
36059 Pick-ups & Expedition '97 thru '09
36060 Super Duty Pick-up, Excursion '99 thru '10
36061 F-150 full-size '04 thru '10
36062 Pinto & Mercury Bobcat '75 thru '80
36066 Probe all models '89 thru '92
Probe '93 thru '97 - *see MAZDA 626 (61042)*
36070 Ranger/Bronco II gas models '83 thru '92
36071 Ford Ranger '93 thru '10 & Mazda Pick-ups '94 thru '09
36074 Taurus & Mercury Sable '86 thru '95
36075 Taurus & Mercury Sable '96 thru '01
36078 Tempo & Mercury Topaz '84 thru '94
36082 Thunderbird/Mercury Cougar '83 thru '88
36086 Thunderbird/Mercury Cougar '89 thru '97
36090 Vans all V8 Econoline models '69 thru '91
36094 Vans full size '92 thru '10
36097 Windstar Mini-van '95 thru '07

GENERAL MOTORS
10360 GM Automatic Transmission Overhaul
38005 Buick Century, Chevrolet Celebrity, Olds Cutlass Ciera & Pontiac 6000 '82 thru '96
38010 Buick Regal, Chevrolet Lumina, Oldsmobile Cutlass Supreme & Pontiac Grand Prix front wheel drive '88 thru '07
38015 Buick Skyhawk, Cadillac Cimarron, Chevrolet Cavalier, Oldsmobile Firenza Pontiac J-2000 & Sunbird '82 thru '94
38016 Chevrolet Cavalier/Pontiac Sunfire '95 thru '05
38017 Chevrolet Cobalt & Pontiac G5 '05 thru '11
38020 Buick Skylark, Chevrolet Citation, Olds Omega, Pontiac Phoenix '80 thru '85
38025 Buick Skylark & Somerset, Olds Achieva, Calais & Pontiac Grand Am '85 thru '98
38026 Chevrolet Malibu, Olds Alero & Cutlass, Pontiac Grand Am '97 thru '03
38027 Chevrolet Malibu '04 thru '10
38030 Cadillac Eldorado & Oldsmobile Toronado '71 thru '85, Seville '80 thru '85, Buick Riviera '79 thru '85
38031 Cadillac Eldorado & Seville '86 thru '91, DeVille & Buick Riviera '86 thru '93, Fleetwood & Olds Toronado '86 thru '92
38032 DeVille '94 thru '05, Seville '92 thru '04
Cadillac DTS '06 thru '10
38035 Chevrolet Lumina APV, Olds Silhouette & Pontiac Trans Sport '90 thru '96
38036 Chevrolet Venture, Olds Silhouette, Pontiac Trans Sport & Montana '97 thru '05
GM Full-size RWD - *see BUICK (19025)*
38040 Chevrolet Equinox '05 thru '09
Pontiac Torrent '06 thru '09
38070 Chevrolet HHR '06 thru '11

GEO
Metro - *see CHEVROLET Sprint (24075)*
Prizm - *see CHEVROLET (24060) or TOYOTA (92036)*
40030 Storm all models '90 thru '93
Tracker - *see SUZUKI Samurai (90010)*

GMC
Vans & Pick-ups - *see CHEVROLET*

HONDA
42010 Accord CVCC all models '76 thru '83
42011 Accord all models '84 thru '89
42012 Accord all models '90 thru '93
42013 Accord all models '94 thru '97
42014 Accord all models '98 thru '02
42015 Accord '03 thru '07
42020 Civic 1200 all models '73 thru '79
42021 Civic 1300 & 1500 CVCC '80 thru '83
42022 Civic 1500 CVCC all models '75 thru '79
42023 Civic all models '84 thru '91
42024 Civic & del Sol '92 thru '95
42025 Civic '96 thru '00, CR-V '97 thru '01, Acura Integra '94 thru '00
42026 Civic '01 thru '10, CR-V '02 thru '09
42035 Odyssey models '99 thru '10
Passport - *see ISUZU Rodeo (47017)*
42037 Honda Pilot '03 thru '07, Acura MDX '01 thru '07
42040 Prelude CVCC all models '79 thru '89

HYUNDAI
43010 Elantra all models '96 thru '10
43015 Excel & Accent all models '86 thru '09
43050 Santa Fe all models '01 thru '06
43055 Sonata all models '99 thru '08

INFINITI
G35 '03 thru '08 - *see NISSAN 350Z (72011)*

ISUZU
Hombre - *see CHEVROLET S-10 (24071)*
47017 Rodeo, Amigo & Honda Passport '89 thru '02
47020 Trooper '84 thru '91, Pick-up '81 thru '93

JAGUAR
49010 XJ6 all 6 cyl models '68 thru '86
49011 XJ6 all models '88 thru '94
49015 XJ12 & XJS all 12 cyl models '72 thru '85

JEEP
50010 Cherokee, Comanche & Wagoneer Limited all models '84 thru '01
50020 CJ all models '49 thru '86
50025 Grand Cherokee all models '93 thru '04
50026 Grand Cherokee '05 thru '09
50029 Grand Wagoneer & Pick-up '72 thru '91
50030 Wrangler all models '87 thru '11
50035 Liberty '02 thru '07

KIA
54050 Optima '01 thru '10
54070 Sephia '94 thru '01, Spectra '00 thru '09, Sportage '05 thru '10

LEXUS
ES 300/330 - *see TOYOTA Camry (92007) (92008)*
RX 330 - *see TOYOTA Highlander (92095)*

LINCOLN
Navigator - *see FORD Pick-up (36059)*
59010 Rear Wheel Drive all models '70 thru '10

MAZDA
61010 GLC (rear wheel drive) '77 thru '83
61011 GLC (front wheel drive) '81 thru '85
61012 Mazda3 '04 thru '11
61015 323 & Protegé '90 thru '03
61016 MX-5 Miata '90 thru '09
61020 MPV all models '89 thru '98
Navajo - *see FORD Explorer (36024)*
61030 Pick-ups '72 thru '93
Pick-ups '94 on - *see Ford (36071)*
61035 RX-7 all models '79 thru '85
61036 RX-7 all models '86 thru '91
61040 626 (rear wheel drive) '79 thru '82
61041 626 & MX-6 (front wheel drive) '83 thru '92
61042 626 '93 thru '01, & MX-6/Ford Probe '93 thru '02
61043 Mazda6 '03 thru '11

MERCEDES-BENZ
63012 123 Series Diesel '76 thru '85
63015 190 Series 4-cyl gas models, '84 thru '88
63020 230, 250 & 280 6 cyl sohc '68 thru '72
63025 280 123 Series gas models '77 thru '81
63030 350 & 450 all models '71 thru '80
63040 C-Class: C230/C240/C280/C320/C350 '01 thru '07

MERCURY
64200 Villager & Nissan Quest '93 thru '01
All other titles, see FORD listing.

MG
66010 MGB Roadster & GT Coupe '62 thru '80
66015 MG Midget & Austin Healey Sprite Roadster '58 thru '80

MINI
67020 Mini '02 thru '11

MITSUBISHI
68020 Cordia, Tredia, Galant, Precis & Mirage '83 thru '93
68030 Eclipse, Eagle Talon & Plymouth Laser '90 thru '94
68031 Eclipse '95 thru '05, Eagle Talon '95 thru '98
68035 Galant '94 thru '10
68040 Pick-up '83 thru '96, Montero '83 thru '93

NISSAN
72010 300ZX all models incl. Turbo '84 thru '89
72011 350Z & Infiniti G35 all models '03 thru '08
72015 Altima all models '93 thru '06
72016 Altima '07 thru '10
72020 Maxima all models '85 thru '92
72021 Maxima all models '93 thru '01
72025 Murano '03 thru '10
72030 Pick-ups '80 thru '97, Pathfinder '87 thru '95
72031 Frontier Pick-up, Xterra, Pathfinder '96 thru '04
72032 Frontier & Xterra '05 thru '11
72040 Pulsar all models '83 thru '86
72050 Sentra all models '82 thru '94
72051 Sentra & 200SX all models '95 thru '06
72060 Stanza all models '82 thru '90
72070 Titan pick-ups '04 thru '10, Armada '05 thru '10

OLDSMOBILE
73015 Cutlass '74 thru '88
For other OLDSMOBILE titles, see BUICK, CHEVROLET or GM listings.

PLYMOUTH
For PLYMOUTH titles, see DODGE.

PONTIAC
79008 Fiero all models '84 thru '88
79018 Firebird V8 models except Turbo '70 thru '81
79019 Firebird all models '82 thru '92
79025 G6 all models '05 thru '09
79040 Mid-size Rear-wheel Drive '70 thru '87
Vibe '03 thru '11 - *see TOYOTA Matrix (92060)*
For other PONTIAC titles, see BUICK, CHEVROLET or GM listings.

PORSCHE
80020 911 Coupe & Targa models '65 thru '89
80025 914 all 4 cyl models '69 thru '76
80030 924 all models incl. Turbo '76 thru '82
80035 944 all models incl. Turbo '83 thru '89

RENAULT
Alliance, Encore - *see AMC (14020)*

SAAB
84010 900 including Turbo '79 thru '88

SATURN
87010 Saturn all S-series models '91 thru '02
87011 Saturn Ion '03 thru '07
87020 Saturn all L-series models '00 thu '04
87040 Saturn VUE '02 thru '07

SUBARU
89002 1100, 1300, 1400 & 1600 '71 thru '79
89003 1600 & 1800 2WD & 4WD '80 thru '94
89100 Legacy models '90 thru '99
89101 Legacy & Forester '00 thru '06

SUZUKI
90010 Samurai/Sidekick/Geo Tracker '86 thru '01

TOYOTA
92005 Camry all models '83 thru '91
92006 Camry all models '92 thru '96
92007 Camry/Avalon/Solara/Lexus ES 300 '97 thru '01
92008 Toyota Camry, Avalon and Solara & Lexus ES 300/330 all models '02 thru '06
92009 Camry '07 thru '11
92015 Celica Rear Wheel Drive '71 thru '85
92020 Celica Front Wheel Drive '86 thru '99
92025 Celica Supra all models '79 thru '92
92030 Corolla all models '75 thru '79
92032 Corolla rear wheel drive models '80 thru '87
92035 Corolla front wheel drive models '84 thru '92
92036 Corolla & Geo Prizm '93 thru '02
92037 Corolla models '03 thru '11
92040 Corolla Tercel all models '80 thru '82
92045 Corona all models '74 thru '82
92050 Cressida all models '78 thru '82
92055 Land Cruiser FJ40/43/45/55 '68 thru '82
92056 Land Cruiser FJ60/62/80/FZJ80 '80 thru '96
92060 Matrix & Pontiac Vibe '03 thru '11
92065 MR2 all models '85 thru '87
92070 Pick-up all models '69 thru '78
92075 Pick-up all models '79 thru '95
92076 Tacoma, 4Runner & T100 '93 thru '04
92077 Tacoma all models '05 thru '09
92078 Tundra '00 thru '06, Sequoia '01 thru '07
92079 4Runner all models '03 thru '09
92080 Previa all models '91 thru '95
92081 Prius '01 thru '08
92082 RAV4 all models '96 thru '10
92085 Tercel all models '87 thru '94
92090 Sienna all models '98 thru '09
92095 Highlander & Lexus RX-330 '99 thru '07

TRIUMPH
94007 Spitfire all models '62 thru '81
94010 TR7 all models '75 thru '81

VW
96008 Beetle & Karmann Ghia '54 thru '79
96009 New Beetle '98 thru '11
96016 Rabbit, Jetta, Scirocco, & Pick-up gas models '75 thru '92 & Convertible '80 thru '92
96017 Golf, GTI & Jetta '93 thru '98, Cabrio '95 thru 02
96018 Golf, GTI & Jetta '98 thru '05
96019 Jetta, Rabbit, GTI & Golf '05 thru '11
96020 Rabbit, Jetta, Pick-up diesel '77 thru '84
96023 Passat '98 thru '05, Audi A4 '96 thru '01
96030 Transporter 1600 all models '68 thru '79
96035 Transporter 1700, 1800, 2000 '72 thru '79
96040 Type 3 1500 & 1600 '63 thru '73
96045 Vanagon air-cooled models '80 thru '83

VOLVO
97010 120, 130 Series & 1800 Sports '61 thru '73
97015 140 Series all models '66 thru '74
97020 240 Series all models '76 thru '93
97040 740 & 760 Series all models '82 thru '88

TECHBOOK MANUALS
10205 Automotive Computer Codes
10206 OBD-II & Electronic Engine Management
10210 Automotive Emissions Control Manual
10215 Fuel Injection Manual, '78 thru '85
10220 Fuel Injection Manual, '86 thru '99
10225 Holley Carburetor Manual
10230 Rochester Carburetor Manual
10240 Weber/Zenith/Stromberg/SU Carburetor
10305 Chevrolet Engine Overhaul Manual
10310 Chrysler Engine Overhaul Manual
10320 Ford Engine Overhaul Manual
10330 GM and Ford Diesel Engine Repair
10333 Engine Performance Manual
10340 Small Engine Repair Manual
10345 Suspension, Steering & Driveline
10355 Ford Automatic Transmission Overhaul
10360 GM Automatic Transmission Overhaul
10405 Automotive Body Repair & Painting
10410 Automotive Brake Manual
10415 Automotive Detaiing Manual
10420 Automotive Electrical Manual
10425 Automotive Heating & Air Conditioning
10430 Automotive Reference Dictionary
10435 Automotive Tools Manual
10440 Used Car Buying Guide
10445 Welding Manual
10450 ATV Basics
10452 Scooters 50cc to 250cc

SPANISH MANUALS
98903 Reparación de Carrocería & Pintura
98904 Manual de Carburador Modelos Holley & Rochester
98905 Códigos Automotrices de la Computadora
98906 OBD-II & Sistemas de Control Electrónico del Motor
98910 Frenos Automotriz
98913 Electricidad Automotriz
98915 Inyección de Combustible '86 al '99
99040 Chevrolet & GMC Camionetas '67 al '87
99041 Chevrolet & GMC Camionetas '88 al '98
99042 Chevrolet Camionetas Cerradas '68 al '95
99043 Chevrolet/GMC Camionetas '94 al '04
99048 Chevrolet/GMC Camionetas '99 al '06
99055 Dodge Caravan/Ply. Voyager '84 al '95
99075 Ford Camionetas y Bronco '80 al '94
99076 Ford F-150 '97 al '09
99077 Ford Camionetas Cerradas '69 al '91
99088 Ford Modelos de Tamaño Mediano '75 al '86
99089 Ford Camionetas Ranger '93 al '10
99091 Ford Taurus & Mercury Sable '86 al '95
99095 GM Modelos de Tamaño Grande '70 al '90
99100 GM Modelos de Tamaño Mediano '70 al '88
99106 Jeep Cherokee, Wagoneer & Comanche '84 al '00
99110 Nissan Camionetas & Pathfinder '80 al '96
99118 Nissan Sentra '82 al '94
99125 Toyota Camionetas y 4-Runner '79 al '95

Over 100 Haynes motorcycle manuals also available

5/22

Haynes North America, Inc., 859 Lawrence Drive, Newbury Park, CA 91320 • (805) 498-6703 • www.haynes.com